# CAREER CHOICES
## *for Students of*

## MATHEMATICS

*by*
## CAREER ASSOCIATES

*Walker and Company*
NEW YORK

First published in the United States of America in 1985 by the Walker Publishing Company, Inc.

Published simultaneously in Canada by John Wiley & Sons Canada, Limited, Rexdale, Ontario.

**Library of Congress Cataloging in Publication Data**
Main entry under title:

Career choices for students of mathematics.

  Bibliography: p.
  1. United States—Occupations. 2. College graduates—Employment—
United States. 3. Vocational guidance—United States. I. Career Associates.
HF5382.5.U5C2555   1985     331.7′023     83-40444
ISBN 0-8027-0796-3
ISBN 0-8027-7250-1 (pbk.)

Printed in the United States of America

10  9  8  7  6  5  4  3  2  1

# Titles in the Series

**Career Choices for Students of:**
Art
Business
Communications and Journalism
Computer Science
Economics
English
History
Mathematics
Political Science and Government
Psychology

**Career Choices for Undergraduates Considering:**
Law
An M.B.A.

## Acknowledgments

We gratefully acknowledge the help of the many people who spent time talking to our research staff about employment opportunities in their fields. This book would not have been possible without their assistance. Our thanks, too, to Catalyst, which has one of the best career libraries in the country in its New York, NY, offices, and to the National Society for Internships and Experiential Education, Raleigh, NC, which provided information on internship opportunities for a variety of professions.

# CAREER ASSOCIATES

# CONTENTS

WHAT'S IN THIS BOOK FOR YOU?

BANKING   1

   Credit Lending   6
   Operations   8
   Systems   10
   Trusts   12
   Interviews   17

THE COMPUTER INDUSTRY   23

   Computer Industry Jobs   24
   Data Processing and Information Systems   32
   Software   41
   Hardware   48
   Training and Teaching   53
   Entrepreneurs   54
   Interviews   55

DEPARTMENT STORE RETAILING   59

   Store Management   63
   Buying   65
   Interviews   71

EDUCATION   75

   Teaching   81
   Interviews   87

INSURANCE   91

   Sales   96
   Actuarial   98
   Underwriting   101
   Interviews   108

## MARKET RESEARCH 111

Research Analysis 115
Interviews 121

## SECURITIES 125

Sales 132
Research 135
Operations 137
Interviews 143

## BIBLIOGRAPHY 149

## INDEX

# WHAT'S IN THIS BOOK FOR YOU?

Recent college graduates, no matter what their major has been, too often discover that there is a dismaying gap between their knowledge and planning and the reality of an actual career. Possibly even more unfortunate is the existence of potentially satisfying careers that graduates do not even know about. Although advice from campus vocational counselors, family, friends, and fellow students can be extremely helpful, there is no substitute for a structured exploration of the various alternatives open to graduates.

The Career Choices Series was created to provide you with the means to conduct such an exploration. It gives you specific, up-to-date information about the entry-level job opportunities in a variety of industries relevant to your degree and highlights opportunities that might otherwise be overlooked. Through its many special features—such as sections on internships, qualifications, and working conditions—the Career Choices Series can help you find out where your interests and abilities lie in order to point your search for an entry-level job in a productive direction. This book cannot find you a job—only you can provide the hard work, persistence, and ingenuity that that requires—but it can save you valuable time and energy. By helping you to narrow the range of your search to careers that are truly suitable for you, this book can help make hunting for a job an exciting adventure rather than a dreary—and sometimes frightening—chore.

The book's easy-to-use format combines general information about each of the industries covered with the hard facts that job-hunters must have. An overall explanation of each industry is followed by authoritative material on the job outlook for entry-level candidates, the competition for the openings that exist, and the new opportunities that may arise from such factors as expansion and technological development. There is a listing of employers by type and by geographic location and a sampling of leading companies by name—by no means all, but enough to give you a good idea of who the employers are.

The section on how to break into the field is not general how-to-get-a-job advice, but rather zeroes in on ways of getting a foot in the door of a particular industry.

You will find the next section, a description of the major functional areas within each industry, especially valuable in making your initial job choice. For example, communications majors aiming for magazine work can evaluate the editorial end, advertising space sales, circulation, or production. Those interested in accounting are shown the differences between management, government, and public accounting. Which of the various areas described offers you the best chance of an entry-level job? What career paths are likely to follow from that position? Will they help you reach your ultimate career goal? The sooner you have a basis to make the decision, the better prepared you can be.

For every industry treated and for the major functional areas within that industry, you'll learn what your duties—both basic and more challenging—are likely to be, what hours you'll work, what your work environment will be, and what range of salary to expect.* What personal and professional qualifications must you have? How can you move up—and to what? This book tells you.

You'll learn how it is possible to overcome the apparent contradiction of the truism, "To get experience you have to have experience." The kinds of extracurricular activities and work experience—summer and/or part-time—that can help you get and perform a job in your chosen area are listed. Internships are another way to get over that hurdle, and specific information is included for each industry. But you should also know that the directories published by the National Society for Internships and Experiential Education (Second Floor, 124 St. Mary's Street, Raleigh, NC 27605) are highly detailed and very useful. They are: *Directory of Undergraduate Internships, Directory of Washington Internships,* and *Directory of Public Service Internships.*

You'll find a list of the books and periodicals you should read to keep up with the latest trends in an industry you are considering, and the names and addresses of professional associations that can be helpful to you—through student chapters, open meetings, and printed information. Finally, interviews with professionals in each field bring you the experiences of people who are actually working in the kinds of jobs you may be aiming for.

---

* Salary figures given are the latest available as the book goes to press.

Although your entry-level job neither guarantees nor locks you into a lifelong career path, the more you know about what is open to you, the better chance you'll have for a rewarding work future. The information in these pages will not only give you a realistic basis for a good start, it will help you immeasurably in deciding what to explore further on your own. So good reading, good hunting, good luck, and the best of good beginnings.

# BANKING

IMAGINE yourself the manager of an operations department, responsible for the global transfer of currencies worth several million dollars. Or a member of the international department, traveling to the Middle East, Africa, or Europe to check on overseas branches. Or managing a loan portfolio for a major multinational corporation, providing its chief financial officer with up-to-date financial information. Banking has become the central nervous system of the world's economy, and today's dynamic banker can be found in front of a desk-top terminal calling up the vast amount of financial data needed to provide an increasing array of new products and services. Today customers want banks to provide more than brokerage services and electronic funds transfers. If you want to be involved in a state-of-the-art business, if you have an entrepreneurial spirit, and, above all, if you are endowed with keen creativity, a career in banking is for you.

The changes in banking are primarily due to the impact of technology. Banking is now a worldwide, 24-hour-a-day business. Automated teller machines, home banking via microcomputers, and office automation have affected every bank employee. But you don't have to be a whiz kid who talks in bits and bytes to get your foot in the door. Every major bank has either a formal training program or professional on-the-job training that includes instruction in the use of the technology. What is most important is your ability to grasp the concept and quickly master the skill.

Banks recruit graduates from a wide variety of majors. In fact, half of all college students entering banking come from a liberal arts background. But don't overlook the traditional financial core courses: business, accounting, marketing, and finance. They will add to your desirability as a job candidate, as will a knowledge of computer science, production management (operations), and interpersonal communications. When a recruiter is having a hard time deciding, it is your interpersonal skills that will count most heavily.

Most banks put their college recruits through a formal training program in which they are taught the methods and practices of the particular institution. Regardless of academic background, all newcomers go into the same melting pot. Students who have taken the financial core courses mentioned will, of course, be more familiar with those subjects during the training program. However, strong analytical skills will enable you to interpret a financial statement, and here an English major who knows how to extract meaning from a careful reading of literature or a history major who knows how to spot a trend or movement in a group of facts will not be at a disadvantage to a finance major.

More and more students entering the field have had the foresight to make themselves knowledgeable about telecommunications to gain an understanding of the newly diverse world of banking. These students have a better chance of getting a job offer than those with a limited, traditional view of the industry.

Many different functional areas exist within banking, and most banks will ask you for which area you prefer to be considered. Commercial and retail banking have recruitment programs in the following functional areas:

- **Credit Lending**
- **Operations**

- **Systems**
- **Trusts**

## Job Outlook

*Job Openings Will Grow:*   Faster than average

*Competition for Jobs:*   Keen
Expect the most competition for positions in credit lending. Expanding opportunities can be found in the operations and systems areas. As new sources for loans become harder to find, operations is being looked to for development of nonfee-based services, such as letters of credit and money transfer services. In systems, the computerization and communications systems needed to deliver customer services are implemented.

*New Job Opportunities:*   Because of industry deregulation, banks are now actively seeking people to work in such diverse areas as mergers and acquisitions; private banking, which serves individuals with high net worth and high incomes; office automation, which develops executive information systems and implements them throughout the bank; product management, which includes the planning, pricing, and marketing of new products and services; and telecommunications, which develops the global communications channels necessary for getting and submitting information.

## Geographic Job Index

Although banks can be found in any city or town, the major money centers are located in New York, NY, Chicago, IL, San Francisco, CA and Boston, MA. Opportunities at the regional or local end of the industry are growing in Dallas, TX, Houston, TX, and other cities in the Southwest.

## Who the Employers Are

COMMERCIAL BANKS (or money-center banks) market their products and services to multinational corporations; to smaller banks, called correspondents; and to individuals, who use checking and loan services.

REGIONAL BANKS provide many of the same services as the larger money-center banks, but on a smaller scale. Their clients are typically locally based small and medium-size businesses.

SAVINGS AND LOAN ASSOCIATIONS offer their customers personal savings accounts and mortgages. However, under new banking legislation, they are allowed to make commercial and business loans.

*Major Employers*

COMMERCIAL BANKS
    Bank of America, San Francisco, CA
    Bankers Trust Company, New York, NY
    Chase Manhattan Bank, New York, NY
    Chemical Bank, New York, NY
    Citibank, New York, NY
    Continental Illinois National Bank, Chicago, IL
    First National Bank of Boston, Boston, MA
    First National Bank of Chicago, Chicago, IL
    Manufacturers Hanover Trust Company, New York, NY
    Security Pacific National Bank, Los Angeles, CA

REGIONAL BANKS
    First Bank System, Minneapolis, MN
    Mellon Bank, Pittsburgh, PA
    Mercantile Bank, St. Louis, MO
    NCNB National Bank, Charlotte, NC
    Ranier National Bank, Seattle, WA
    Republic Bank Dallas, Dallas, TX
    Wachovia Bank & Trust Company, Winston-Salem, NC

## How to Break into the Field

Most banks have formal on-campus recruitment programs through which they hire most of their trainees. They frequently recruit separately for each major functional area: credit lending, operations, systems, and trusts. Be sure to check schedules carefully to ensure an appointment in your area of interest.

Before the interview, do your homework. Learn all you can about the internal workings of the area for which you plan to interview. If your field of interest is not represented, select the next most appropriate area and ask the recruiter to forward your résumé to the proper section. Also, learn something about the bank itself. Different banks have different personalities. Some are aggressive, others more traditional and conservative. Try to interview with banks that have a corporate identity compatible with your own personal identity.

Landing a summer internship is another pathway to a full-time position. Most major banks have internship programs, although they are usually limited to graduate students. Recruitment for the internships is usually done through campus visits. Check with your placement office for details.

If your college does not have a formal placement office, or if the bank to which you wish to apply is not recruiting on your campus, send a well-written letter, accompanied by a résumé, to the bank's director of college recruitment. Follow up your letter with a phone call.

Whether you have an on-campus interview or are writing directly to the college recruitment department, never pass up help from anyone who knows someone at the bank. A well-placed word can be invaluable.

## International Job Opportunities

At a large commercial bank, and even at some regional banks, overseas work is possible. International department lending officers may be assigned to work abroad for a period of three to five years, or may be required to travel abroad frequently. Corporate department staffs that handle U.S. multinational corporations also do quite a bit of business overseas.

Most banks try to staff their overseas branches with local citizens. Only the higher-level managerial jobs may be filled by Americans. Specialized positions in areas such as investment banking, joint ventures, and trade go to M.B.A.s or other experienced personnel. Fluency in a foreign language is helpful but not essential, because most banks have contracts with language schools to provide training as necessary.

# CREDIT LENDING

This is the most visible area of banking, the area that involves the traditional bank-client relationship that almost everyone associates with the industry. However, this aspect of banking is more than just extending credit or offering interest-bearing accounts to clients. In consumer banking, a lending officer assesses the creditworthiness of individuals. In commercial banking, a lending officer evaluates the financial status of corporations or nonprofit organizations; performs industry surveys, analyzing a particular industry to determine if backing a firm in that area is a good loan risk; makes production forecasts to see if a borrowing firm's available resources will meet production requirements; predicts how a loan would affect the bank's cash flow positively or negatively; or handles corporate overdrafts, contacting corporate customers whose payments are late.

To start out you will go on customer calls with experienced loan officers and be responsible for taking notes and writing a report on the customer and the loan review—not as a participant, but as an observer. You may be called on to research new business prospects, making cold calls to prospects in a given territory or industry. At a smaller bank, your responsibilities will be broader and you will actually make decisions on modest loans quite early.

## Qualifications

*Personal:*   Strong analytical skills. Ability to conceptualize. An affinity for quantitative problems. Strong negotiation skills. Extremely good interpersonal skills.

*Professional:* Ability to analyze data and financial statements and do creative financial planning. Familiarity with bank products and services. Ability to present clearly written reports.

## Career Paths

| LEVEL | JOB TITLE | EXPERIENCE NEEDED |
| --- | --- | --- |
| Entry | Trainee | College degree |
| 2 | Assistant Loan Officer | 1-2 years |
| 3 | Loan Officer/Branch manager | 3-5 years |
| 4 | Loan manager | 7+ years |

## Job Responsibilities

### Entry Level

**THE BASICS:** Training will consist of both classroom instruction in such areas as finance, accounting, and credit analysis, and actual account work, helping lending officers make judgments about existing or potential bank relationships.

**MORE CHALLENGING DUTIES:** Upon completion of training, you will be assigned to a line lending area, attend advanced banking seminars, and have the opportunity to meet with customers.

### Moving Up

Your advancement will depend on your ability to establish advantageous client relationships, to close lucrative loan deals successfully, and to know when not to approve a loan. As you advance, the loan review process will become more complex and involve significantly more money. You can measure your success by your approval authority—how big a loan you are authorized to approve without going to a higher level of management.

# OPERATIONS

The most successful banks anticipate and satisfy all their customers' financial needs. Operations occupies a front-row seat in the banking industry because it has bankwide responsibility for providing customers with nonfee-based (nonloan) services—letters of credit, money transfers, and foreign exchange—services of increased importance because banks can no longer make the profits they once did by lending money to customers. The operations department is usually the largest department of a commercial bank. The Chase Manhattan Bank operations department, for example, has more than 4,000 employees. Graduates are employed in supervisory positions, managing the clerical staff, with responsibility for setting up assignments and time schedules, evaluating performance, making sure work is done properly, training new employees, and authorizing salary increases. Work in operations also involves troubleshooting for customers, solving their account problems, for example, by tracing a money transfer that was never credited.

## Qualifications

*Personal:*   Ability to meet deadlines.  Ability to perform under pressure. Ability to get along with many different types of people.

*Professional:*   Ability to understand and follow through on complex instructions. Familiarity with concepts of computer science or a related discipline. Knowledge of fee-based services and products.

## Career Paths

| LEVEL | JOB TITLE | EXPERIENCE NEEDED |
|-------|-----------|-------------------|
| Entry | Operations Trainee | College degree |
| 2 | Supervisor | 18 months |
| 3 | Department manager | 3-5 years |
| 4 | Division manager | 6+ years |

## Job Responsibilities

### Entry Level

**THE BASICS:** You begin your career in operations either in a formal training program, or, more likely, on the job. You will be an operations trainee for about 18 months, learning by rotating among the various departments that handle fee-based services.

**MORE CHALLENGING DUTIES:** After the training period, you will be assigned to a department or a staff area such as financial management or budget coordination and will learn about a single product or area in depth.

### Moving Up

Your progress will depend on your ability to improve the overall productivity of your department or area, to motivate your staff, to stay within your budget, and to complete transactions efficiently and accurately. Because operations is not exclusively devoted to production management, for further advancement you will need to learn about product development, marketing, and systems functions. Those who move into these areas often accompany loan officers on customer calls, offering the technical advice that will help clinch a deal or presenting a plan to customize an existing product to meet the client's expanding needs.

With hard work and diligence you can acquire the knowledge and expertise that will enable you to move almost anywhere in the bank organization. Operations managers can move into marketing positions, the systems areas, or perhaps relocate (even overseas) to manage a branch bank.

# SYSTEMS

The systems area is now involved in every banking decision from credit lending to recruitment. Most large commercial banks have both a central systems area and separate decentralized systems units that service the major components of the organization. Systems is responsible for developing, implementing, and maintaining automated programs for clients and for in-house use; for selecting hardware, writing software, and consulting with the user-client when special programs must be developed. In addition, systems staffers must keep up with the latest developments in technological applications and services.

## Qualifications

*Personal:*   Ability to think in analytical terms. Ease in working with abstract models.

*Professional:*   Quantitative skills. Familiarity with the business applications of software and hardware. Ability to convert technical language and concepts into familiar and understandable terms.

## Career Paths

| LEVEL | JOB TITLE | EXPERIENCE NEEDED |
|-------|-----------|-------------------|
| Entry | Systems trainee | College degree |
| 2 | Systems analyst | 2 years |
| 3 | Systems consultant | 3 years |
| 4 | Senior systems consultant | 5 years |

## Job Responsibilities

### Entry Level

**THE BASICS:** Either in a structured training program or through on-the-job-training, you will become familiar with the bank's hardware and software and how they are used. Depending on your background, you may become a programmer, or you may be placed on a systems team project, refining the use of current equipment or developing systems for as yet unmet needs.

**MORE CHALLENGING DUTIES:** Applying your skills to more difficult or specialized projects.

### Moving Up

If you demonstrate interpersonal skills as well as technical ability, you could become a project manager, overseeing a team of systems people working on the development and implementation of a specific systems capability, such as a new internal telephone switching system or software for an executive work station, which could include features such as electronic mail and word processing.

The potential for a talented systems person is excellent. You could end up managing an operations or office automation department, developing and installing new systems, or becoming a systems consultant for overseas branches. Successful systems personnel can move into any department in the bank.

# TRUSTS

The trust department manages and invests money, property, or other assets owned by a client. The pension plans of large corporations and other organizations often use trusts, as do individuals with large assets. Many estates are also managed in trust by the provisions of a will. Like the credit department, this department deals closely and extensively with clients. The training program is similar to that in other areas of banking, but in general advancement is slower and requires more experience.

## Qualifications

*Personal:* A straightforward manner. Accuracy. Good with numbers. Patience in dealing with people. Confidence.

*Professional:* Strong analytical ability. Good business judgment. Ability to apply financial theory to practical problems.

## Career Paths

| LEVEL | JOB TITLE | EXPERIENCE NEEDED |
|-------|-----------|-------------------|
| Entry | Trainee | College degree |
| 2 | Assistant trust officer | 1-3 years |
| 3 | Trust officer | 4-6 years |
| 4 | Senior trust officer | 10+ years |

## Job Responsibilities

## Entry Level

**THE BASICS:** Developing familiarity with bank policies and procedures.

**MORE CHALLENGING POSITIONS:** Researching investments, real estate, or the overall economy to assist superiors. Some contact with clients.

## Moving Up

Showing sound judgment and an ability to work independently will garner an assignment to manage some of the smaller trust funds. Moving up also depends on your ability to attract new customers to the bank, as well as to keep present clients satisfied. As you advance you will become responsible for handling more and more money. Top-level trust officers are expected not only to bring in substantial new business and to handle the largest accounts, but also to manage and support lower-level employees.

# ADDITIONAL INFORMATION

## Salaries

Salaries vary according to the size of the bank. The following figures are taken from Robert Half International's 1984 survey:

Installment loans/assistant manager: $18,000 to $22,000 (small bank); $21,000 to $27,000 (medium-size bank); $23,500 to $28,500 (large bank).

Commercial loans/branch manager: $22,000 to $28,000 (small); $24,000 to $31,000 (medium); $26,000 to $31,000 (large).

Senior loan officer:   $28,000 to $32,000 (small); $33,000 to $37,000 (medium); $33,000 to $50,000 (large).

Mortgage loans:   $23,500 to $32,000 (small); $28,000 to $36,000 (medium); $32,000 to $41,000 (large).

Operations officer:   $17,000 to $21,000 (small); $22,000 to $29,000 (medium); $24,000 to $31,000 (large).

Trust officer:   $22,000 to $29,500 (small); $23,000 to $30,000 (medium); $27,500 to $40,000 (large).

## Working Conditions

*Hours:*  The credit trainee rarely sees daylight, because long hours and weekend work are often required to get through the training program. After training, normal hours will be whatever it takes to get the job done (nine to five plus). The hours in operations are different because it is a 24-hour-a-day shop. Night shifts and weekend work may be unavoidable, especially for less experienced employees. Systems staffers may also work on a 24-hour clock; the hours are longest when new systems are being installed and deadlines must be met.

*Environment:*   Lending officers get the choicest locations in the bank; because their job is customer-oriented, the surroundings are usually plush and pleasant. The operations and systems departments take a 360-degree turn from the lending department; the workspace is strictly functional, with few amenities.

*Workstyle:*   In credit, much time is spent researching facts and figures about existing and prospective clients, which could take you from the bank library to the client's headquarters. The rest of your time will largely be spent in conference with senior lending officers. Operations and systems work is desk work. Managers walk the area, talking with the staff and lending assistance. In both departments, senior people may meet occasionally with systems consultants.

*Travel:* Travel is rare for entry-level employees in any bank. Later, however, lending officers in consumer banking might travel throughout their state. In commercial banking, research could take a lending officer to major cities throughout the country. If you are assigned to the international department in credit, operations, or systems, you might be sent to overseas branches.

## Extracurricular Activities/Work Experience

Experience as a cashier/teller

Clerical experience

Financial officer/treasurer in campus organizations

## Internships

Many banks—savings and loan associations and consumer and commercial banks—are willing to take interns, especially in summer programs. Interns are paid, and the experience may result in a job offer after graduation. Your campus placement office is the best source of information regarding these programs. If your school does not have a placement office, contact the college recruitment director at banks that interest you for details.

## Recommended Reading

**BOOKS**

*All You Need to Know About Banks* by John Cook and Robert Wood, Bantam Books: 1983

*The Bankers* by Martin Mayer, Ballantine Books: 1980

*In Banks We Trust* by Penny Lernoux, Doubleday & Company: 1984

*Money: Bank of the Eighties* by Dimitris Chorafas, Petrocelli: 1981

*Money and Banking* by Richard W. Lindholm, Littlefield, Adams & Company: 1969

*The New Age of Banking* by George Sterne, Profit Ideas: 1981

*Polk's World Bank Directory,* R. L. Polk and Company (semiannual directory listing banks by city, state, and foreign country)

*Your Career in Banking,* American Bankers Association: 1980

**PERIODICALS**

*ABA Banking Journal* (monthly), 345 Hudson Street, New York, NY 10014

*American Banker* (daily), One State Street Plaza, New York, NY 10004

*The Banker's Magazine* (bimonthly), Warren, Gorham, and Lamont, Inc., 210 South Street, Boston, MA 02111

*Bank News* (monthly), 912 Baltimore Avenue, Kansas City, MO 64105

## Professional Associations

American Bankers Association
1120 Connecticut Avenue, N.W.
Washington, DC 20036

Consumer Bankers Association
1725 K Street, N.W.
Washington, DC 20006

National Association of Bank Women
111 East Wacker Drive
Chicago, IL 60601

United States League of Savings Associations
111 East Wacker Drive
Chicago, IL 60601

# INTERVIEWS

**Louise D'Imperio, Age 22**
**Operations Analyst**
**Chase Manhattan Bank, New York, NY**

My association with Chase began while I was a student at Villanova University. During summer breaks, I was a member of the apprenticeship management training program, which places undergraduates in operations. The program provides quality relief for full-time employees who take vacations.

I worked in the interbank compensation department, which is responsible for the settlement of funds transfer errors. I worked in the staff support section, which supports the production line. I began by doing simple clerical functions, but later became involved in numbers crunching for production tracking reports. In my final summer, I was an inquiry clerk. My responsibility was to take customer and other bank questions over the phone and via telex and inform the individual of the outcome of the compensation case or reconcile any errors made in settling the case.

In that department I started from the ground up. After three summers, I really knew how a case was initiated and processed, and I had a knowledge of the problems that can arise. But after I graduated in May 1983 I wanted a job outside of bank operations. I have a B.S. in business administration with a concentration in marketing. I wanted a marketing-oriented job and I wanted to be involved in product positioning.

Because I had contacts at Chase, I was able to bypass the normal channels that graduates go through. I looked outside of banking, and mailed résumés to various departments at Chase. Among others, I got a response from Chase international operations and systems.

I chose the position in international operations and systems because I felt that a job in office automation would open up an interesting career path. I knew very little about the field of office automation, but was very interested in it. I work in a division that is concerned with office automation in the international section—more specifically, smaller Chase branches abroad. I'm involved in the marketing and support function of the division, which markets office automation products internally. We want to increase the productivity of individual branches, and we want to increase the use of our products. Our work involves training, consultation, and the development of customized software.

The brunt of my work is project-oriented. Right now I'm working on a project that examines what office automation may do for one of Chase's small subsidiaries. I also edit an office automation newsletter, which takes up about 40 percent of my time, and have written documentation for some of the software developed by our group.

I knew nothing about office automation when I started this job; I actually thought that it involved only word processing. Office automation goes way beyond word processing to include a variety of technologies. My background in operations was not a requirement for this job, but it has made it easier to view the workings of the bank. It also showed me how much I still have to learn about banking. I enjoy my job and I like being involved with technologies that have a definite impact on productivity.

**Jayne Geisler, Age 32**
**Vice President, Market and Financial Planning**
**Chemical Bank, New York, NY**

After receiving a B.A. degree in mathematics and French in 1973 from the State University College of New York at Potsdam, I entered the M.S. teaching program at Boston College, which combined coursework with a part-time teaching position in high school mathematics. Finding teaching unchallenging and realizing my abilities would be better utilized in the business environment, I entered banking, an industry where I felt I could capitalize on my quantitative background

I joined Chemical Bank in 1974 as a financial analyst in the finance, then control, division. My responsiblities included cost accounting and financial management reporting for the consumer banking and upstate regions of the metropolitan (New York) division. Specifically this consisted of preparing, analyzing, and monitoring the financial performance of these business segments against budget and prior years, plus the development of unit and product costs of various banking services. The work was entirely hands-on, with no formal training program, and provided me with a broad understanding of the mechanics of the banking industry.

In 1977 I transferred to the controller's area of the metropolitan division where my duties expanded to include perormance reporting and analysis for the commercial as well as consumer lending areas of the division, acting as a liaison with these areas, plus coordinating their annual budgets. In addition, I was charged with designing and implementing a management information system for evaluating the financial performance of these business segments against budget.

Since 1975 I had been working toward my M.B.A. in finance at night from New York University. Coming from a nonbusiness educational background, I felt that it was apparent that an M.B.A. was necessary to enhance my professional development and my future career goals. It provided me with an understanding of the interrelationships among the key business ingredients—finance, economics, marketing, management, and accounting—which I thought necessary to be more effective in my job. As a result, I am of the opinion that an M.B.A. is an excellent degree for enhancing one's background, especially for those with a liberal arts education. However, I strongly believe that business school is more meaningful and relevant to those who have had prior work experience, as there exists a context in which to augment the course of study.

Upon completion of my M.B.A. in February 1979 I entered the bank's commercial credit training program in order to be a part of the bank's basic business—lending—and to round out my banking experience. I was assigned to the district specializing in the garment/textile/entertainment industries. Handling a portfolio of small business and middle market customers was a challenge. I analyzed and determined credit needs, structured deals, and provided cash management servicing.

Late in 1980 I was asked to join the division's strategic planning unit, which was then undergoing expansion. After a little more than a year as deputy department head, I was promoted to director of the unit, which is my current position. Planning has become increasingly important due to the deregulation of the banking industry. "What do we do now? Where do we want to be in five years? What new products/services should we offer?" These are just some of the challenges facing us as we anticipate the changes in banking law and the movements of our competition. In view of this changing environment created by deregulation, I began working toward a law degree to further supplement my background and experience.

Banking is experiencing tremendous growth and change—it's a whole new ballgame—evolving into a fully integrated financial services industry. The competition not only includes banking institutions, but has expanded to comprise brokerage and investment houses, retailers, high-tech companies, conglomerates, and so forth. As a result, those individuals seeking to enter the industry

will need to be sales-oriented and well-rounded in financial services. Banking, finance, and credit will provide the basis, but
securities, insurance, and other financial services will play key
parts in the banking financial supermarket.

# THE COMPUTER INDUSTRY

THE U.S. Department of Labor estimates that the total number of computer occupations will grow from 1,455,000 in 1980 to 2,140,000 in 1990, an increase of 47 percent. The number of programmers will jump 47 percent from 341,000 in 1980 to 500,000 by 1990. Jobs as systems analysts will show a 65 percent gain, climbing from 243,000 to 400,000. And computer service technicians will almost double in number, from 83,000 to 160,000. Clearly the computer industry is a growth industry that shows no signs of slowing down in the next decade.

Computer personnel are involved in virtually every aspect of American business and industry. Programmers work on such diverse projects as writing specialized accounting programs for in-house use in large insurance companies to developing sophisticated interfaces between computer-aided design systems and robots on the factory assembly line. Systems analysts design software for manufacturers, supermarket chains, banks, oil companies, government, computer service companies, and medical researchers, among many others. Some computer professionals work on highly technical applications in computer hardware, while others concentrate on creating software that is so "friendly" virtually anyone can use it with minimal training. In general, wherever computers are used, job opportunities exist for computer professionals.

So many avenues are open to computer science graduates that it's hard to choose and hard to know where to begin. Your decision will be considerably easier if you know what the jobs are and what they entail, and in this introductory section you will get an idea of what the computer industry has to offer.

One thing you can offer the computer industry is the mastery of a very special language—English. Although your knowledge of computer languages is of course important, the skill most lacking in the industry today is an excellent command of spoken and written English. One of the major causes of company failures has been the inability of staff to explain to customers the capabilities of their products and how to use them. Computer science graduates who are also excellent communicators are and will continue to be in high demand.

# COMPUTER INDUSTRY JOBS

### Progamming

Programmers write the codes that instruct the computer. They test their programs, debug them of any errors, and sometimes write the accompanying documentation that tells others how to apply the program. After the systems analyst designs a program for a given need, the programmer must be able to follow the directions and write a program that will perform the prescribed task. The systems analyst will detail everything that a program should do, but it is the programmer's job to figure out how the computer should perform these tasks and in what order.

Programming divides into two main categories: systems programming and applications programming. While applications programmers write codes that enable the computer to perform specific tasks, the systems programmer prepares the computer to understand the language that the applications programmer will be using and tells the computer what peripheral equipment—printers, automatic teller machines, etc.—it will be controlling.

The systems programmer primes the computer for the task at hand, using low-level languages, allocating enough memory for the prescribed functions, and setting up the priorities for those functions. When new hardware is purchased, systems programmers integrate it into the existing systems, and when computers "go down," they diagnose the problem and fix it.

Systems programming is considered more technically oriented than applications programming, and salaries are generally higher, reflecting a greater need.

Applications programmers write the programs that tell the computer how to perform specific tasks. Commercial applications programmers, using high-level languages, work on computer applications to implement business functions. These range from automatic reordering to design and manufacture, scientific and engineering applications, and some heavily mathematical commercial applications.

## Qualifications

*Personal:* A logical mind. Ability to concentrate on routine. Good memory. Keen eye for detail.

*Professional:* Systems: Degree in computer science or one year's experience with an assembly language. Knowledge of interfacing and modifying communications or data base software helpful.

Commercial applications: Working knowledge of COBOL, some BASIC, RPG II, and Pascal helpful. Exposure to large operating systems, data base management, direct access techniques, remote processing, virtual systems, CRT drivers and data base handlers gives an edge.

Scientific applications: Experience with FORTRAN or assembly language. Undergraduate degree in mathematics or engineering preferred. Exposure to large-scale hardware or mini/micro systems a plus.

## Job Responsibilities

**APPLICATIONS:**　Write detailed program design. Code, test, and debug programs. Write documentation for programs.

**SYSTEMS:**　Devise and maintain operating systems. Maintain monitors, data base packages, compilers, assemblers, and utility programs. Plan and evaluate hardware and software for acquisition and modify new equipment to fit the needs of the company.

## Moving Up

Commonly, a programmer trainee will work as part of a team under the supervision of a lead programmer. With added experience, programmers sometimes head up their own projects within the team while lending their efforts to projects led by other team members. All programmers work closely with the systems analysts who design the programs that they create.

A good programmer will find it relatively easy to move up the ranks to senior programmer, then lead programmer, or to move over to systems analysis. In some companies, employees called programmer analysts perform some of the duties of both positions. Where they rank in the hierarchy of computer personnel varies from company to company.

It is wise to keep up with the technology. New developments in self-programming software will cut down the need for applications programmers eventually, and today programs that can be constructed simply by building on packaged software already exist. The trend is toward more technologically oriented jobs as the microcomputer makes it easier for the general user to perform tasks that previously were the exclusive domain of computer personnel.

## Systems Analysis

Although programmers deal primarily with machines, systems analysts deal with the people who use the machines, adapting computer hardware and software to end-users' needs. Listening is perhaps the most important part of the job. After understanding just what the customer wants, the systems analyst examines the computer equipment the company already has and determines how this existing hardware can be adapted to the task. He or she then will determine what interfaces will have to be designed, what codes employees will have to learn, what additional equipment must be purchased, and what programs have to be written. The systems analyst then designs the program, which a programmer will then code, and oversees the entire project.

Because systems analysts are responsible for selecting the best equipment for the job, they must keep up with the latest developments in computer technology. They must also be good communicators. Some background in business (or advanced math or engineering for scientific applications) is helpful.

## Qualifications

*Personal:*   Good communications skills. Personable. Able to get along well with many different kinds of people.

*Professional:*   Programming experience essential. Some background in business.

## Job Responsibilities

Deal with clients. Identify client's needs and design systems to meet them. Supervise implementation of the system and coding of programs. Assess finished system.

## Moving Up

Systems analysts usually move up to a management track or to marketing support positions, advancing from project leadership to technical services management, systems and programming management, or operations management. M.B.A.s are preferred candidates for promotions, and the trend today is for systems analysts to return to school for advanced business degrees in order to be considered for promotion into management.

## Data Base Management

The data base specialists monitor all data in the system for redundancies, backups, and ease of access. They control the flow of data, channel it into the right files, and provide for the recovery of files in case of emergency. Another important aspect of the data base specialist's job is security. He or she is one of several computer professionals who must be on the watch for such activities as embezzlement, vandalism (the destruction of portions of the memory by disgruntled employees) and other computer crime.

## Qualifications

*Personal:*   Attention to detail. Personal integrity.

*Professional:*   Familiarity with systems programming. At least five to ten years' experience in data processing. Some knowledge of a major CODASYL or hierarchical data base package, distributive systems or communications, and relational data bases.

### Job Responsibilities

Monitoring the company's data resources. Consulting with company planners when new hardware is being purchased. Conferring with department managers when a breach in system security is detected.

## Moving Up

Data base specialists usually move along a management track, culminating in the position of corporate data base administrator, with responsibility for the company's entire data resource.

### Product Support

Technical support representatives, sometimes called systems engineers, are employed by computer hardware and software manufacturers. They are problem solvers, answering questions and working out any difficulties the customer may be having with the system. Like a systems analyst, a technical support representative will consult with the potential customer before anything is purchased, analyzing the needs of the company, putting together feasibility studies, writing sales proposals, and demonstrating the product.

   The demand for technical support representatives continues to increase, particularly in the areas of visual display units, data base management systems, software and custom systems, and mini and micro systems.

Product support representatives work closely with sales representatives, but are paid a salary and do not receive commissions.

### Qualifications

*Personal:*Diligence. Efficiency. Problem-solving ability. Good interpersonal skills.

*Professional:*Preferably a four-year degree in computer science, engineering, or some other technical discipline. Very good written and oral communications skills.

### Job Responsibilities

Keeping the customer happy and comfortable with the computer. Making sure the product will meet all the buyer's expectations.

## Moving Up

People in product support usually move on to positions in marketing, technical services management, or consulting.

### Sales and Marketing

In the computer industry, it's not uncommon for sales personnel to meet with potential customers for months, making sure that they are selling the right equipment for the intended job. Sales personnel must know the products they are selling and keep up with the latest technology, attending company-run seminars. Programmers and systems analysts who move into this area naturally have the advantage of a technical background and computer experience, which helps them to understand and explain the newest products and services. An added incentive for moving over into sales is the potential for high earnings.

Marketing personnel design the marketing strategy of a product or service, determining at whom their efforts should be aimed, and how potential customers should be approached. They decide whether to advertise and where. With the boom in mini and micro systems and packaged software, the trend toward retail sales is expected to continue as more people purchase their first home computers and present computer owners upgrade their systems. As a result, hardware and software manufacturers will be hiring additional marketing personnel to specialize in retail trade.

## Qualifications

*Personal:*   Good communications skills. Confidence. Assertive (not aggressive) personality. Ease in dealing with people.

*Professional:*   Four-year degree in business helpful, although many sales representatives have only technical backgrounds. Good selling skills.

## Job Responsibilities

Selling company's services. Servicing customers' accounts after sale is made.

# Moving Up

People in sales and marketing who achieve outstanding sales and reach previously untouched markets are excellent candidates for marketing management. Management positions in sales and marketing are generally the most lucrative of all jobs in the computer industry.

## EDP Auditor

Department heads can monitor the systems under their supervision, but someone is needed to evaluate the computer functions of the entire company, and this is the job of the EDP (electronic data processing) auditor. It is predicted that billions of dollars are lost each year to computer inefficiency. EDP auditors evaluate a company's systems and operational procedures and recommend improvements to top management.

Some EDP auditors work within corporations in the finance or auditing departments while others work for outside accounting firms. After a year's experience, an EDP auditor can earn between $23,000 and $25,000. After four years on the job, an EDP auditor can earn in the neighborhood of $44,000 annually.

## Qualifications

*Personal:*   Honesty, impartiality, and integrity. Diligence. Investigative ability. Abstract thinking. Good communications skills.

*Professional:* Broad background in programming and systems. Some knowledge of many different computer systems, languages, and applications. Some formal training in business and accounting. Certification as public accountant helpful.

## Job Responsibilities

Conduct detailed evaluations of extant and proposed systems. Present findings to management with specific recommendations for improved accuracy, procedures, and security.

## Moving Up

Because EDP auditors have a direct line to top management, EDP auditing is considered an excellent job for people who wish to move up into consulting or corporate management, and for this reason many programmers and systems analysts seek out auditing positions.

## Documentation Specialist

After a programmer finishes coding a new program, a technical manual is written in ordinary English, explaining what the program does, how it does it, and how to use it. In some cases programmers must write their own documentation, but more frequently a documentation specialist—also called a technical writer—is responsible for it. Basically, the documentation specialist translates technology into plain, comprehensible English.

Documentation specialists consult with the systems and programming staff and with the end-user, ensuring that the customer is using the product the way it was intended to be used. Documentation personnel are sometimes called on to write promotional brochures and advertising copy. They also participate in systems specifications and design.

Many documentation specialists are self-employed and contract for assignments from companies on a free-lance basis.

## Qualifications

*Personal:* Ability to draw out information from programmers. Attention to detail. Strong organizational skills. Logical thinking.

*Professional:*   Good writing skills. Some technical background highly recommended.

### Job Responsibilities

Making computer technology accessible to people with no technical background. Essentially, documentation personnel are translators.

## Moving Up

The demand for documentation specialists is growing with the expanded use of time-sharing and distributive processing systems, so jobs will be plentiful. Those with some background in advertising can move over into marketing. Otherwise, documentation is considered a separate career track in the computer industry, one that does not lead to management.

# DATA PROCESSING AND INFORMATION SYSTEMS

It goes without saying that business and industry have been revolutionized by the computer. The burden of paperwork, the tedium of record-keeping, and the frustration of trial-and-error industrial testing have been virtually eliminated because computers are able to digest huge volumes of data and turn it into usable information. These accomplishments are the work of the systems analysts and programmers who design and create the software that enable computers to meet the specific needs of the end-user.

In the coming years, the demand for computer professionals will be high as the need for better and faster access to information increases. As computer technology continues to develop systems for business and industry, companies will be hiring people to adapt these new systems to their particular needs.

### Job Outlook

*Job Openings Will Grow:*   Faster than average

*Competition for Jobs:*   Some

*New Job Opportunities:* According to the Department of Labor's Bureau of Labor Statistics, there will be 2,140,000 computer-related jobs in the United States by 1990—a 47 percent increase over 1980. Needless to say, data processing and information systems are growth industries, and opportunities are plentiful for recent graduates with some experience in programming. Computer services, which include service companies, hospitals, and educational institutions, will provide the greatest number of jobs in programming, according to the Bureau of Labor Statistics, showing a 109 percent increase in the number of jobs in 1990 over 1980. Other fields that will show dramatic increases in computer-related jobs are manufacturing, government, finance/insurance/real estate, and wholesale and retail trade.

## How to Break into the Field

The demand for programmers and systems analysts is growing every day, and many employers prefer candidates who took double majors, combining computer science with accounting, business administration, or math. A working knowledge of one or more computer languages, preferably COBOL, PL/I, BAL, RPG II, or UNIX-C, is a must if you want to be considered for an entry-level position.

Although programmer positions are plentiful now, they will start to dwindle with the development of high-level self-programming languages. Applications programming is a fine place to start for your first job, but as you gain experience you should look toward areas that will grow with you, such as advanced technical training in systems design, data network communications, computer engineering, marketing, and management.

Don't wait for the jobs to come to you through your college placement office. Determine what your interests are and in what areas you would most like to work, then write to the companies in that area. Write directly to the head of the data processing department or even to the president of the company. This kind of initiative tells the employer a great deal about you, and in most cases your résumé will get more attention when it's addressed directly to the person who actually does the hiring.

## Geographic Job Index

Jobs can be found all over the country. Service companies tend to cluster around larger cities, close to the small to medium-size firms they serve, and manufacturers, wholesalers, and retailers also tend to be situated near or close to the larger urban centers. Federal government jobs will be most numerous in Washington, DC, and in those large cities where government agencies have branches. The larger the state or municipal government, the more jobs it will have for computer professionals. The highest concentration of jobs in the insurance industry and financial institutions is in the Northeast.

## Who the Employers Are

SERVICE COMPANIES are the largest employers within the service category, which employs about one-third of all computer personnel. Computer programming service companies offer all the benefits of being on a large mainframe system to small and medium-size companies who could not afford to buy their own computers and hire their own programmers. Typically, they are modest companies, but some customers are as large as the Department of Defense. Service companies do batch processing, computerizing their customers' more cumbersome paperwork, and also provide information services over the telephone lines for an hourly fee.

They are often hired to create custom software to fit the particular needs of a customer company with a new computer. Their systems analysts usually have some training in the areas for which they design software.

Service companies are a growth industry and are projected to continue to grow over the next ten years, adding many new jobs to the market. Employers are willing to consider new graduates, particularly those who have some hands-on experience with operational systems. Double majors are preferred, especially accounting/computer science, as is exposure to the more advanced computer languages, such as BAL or AA. Most service companies do hire programmer trainees. However, many like to hire from within their own ranks, promoting loyal employees and training them in-house. Your stiffest competition may come from data-entry operators reaching for promotions. But the demand for programmers is so great, that you need not worry about a shortage of jobs.

MANUFACTURING FIRMS are beginning to take advantage of emerging technology, and more computer professionals will be needed to create and adapt software to the needs of industry. In the next ten years, process control, quality control, business forecasts, and management information functions will be further computerized. CAD (computer-aided design) systems have made it easier for engineers to design and test new products, and CAM (computer-aided manufacture) systems are being developed to control the total operation of the factory, from designing the right tools to controlling the robots on the assembly line to monitoring supply levels and ordering materials as needed. The big stumbling block will be linking CAD systems to CAM systems, and this is the one area where manufacturing firms will need the most computer personnel.

Another goal of industry will be the implementation of the "flexible factory," in which a single assembly line is able to produce multiple variations of a product, allowing profitable, low-volume output of a customized item. The Deere and Company plant in Waterloo, IA, is such a flexible factory currently in operation, capable of making more than 5000 tractor configurations on their assembly line. On a conventional assembly line, it might take a month or more to retool for a new variation of the product, but on the flexible assembly line each product can be different from the one before it. The specifications are simply punched in and the robots do the rest.

Relatively few American manufacturers have flexible factories right now, but in the next decade more will be built, particularly by those companies producing electronics components and heavy products such as automobiles. Eventually factories will have all operations working off the same mainframe, from office accounting to quality control at the end of the assembly line. Linking these functions will require a great many programmers and systems analysts. Many large manufacturers are now beginning the process of modernization, and are hiring their own computer personnel to write the programs they will need.

# GOVERNMENT

GOVERNMENT at the federal level is the single biggest end-user in the world, maintaining its own computers as well as subcontracting much of its work to service companies. One-fourth of the Social Security Administration's budget goes into data systems, and it is projected that by 1985 the federal government will have 29,000 computers.

The need for qualified personnel to service these computers will continue to grow.

The demand for computer professionals is even greater in state and local government. Single function systems such as payroll, benefits, revenue collection, and disbursement are being consolidated. New jobs for programmers and systems analysts will become available as other functions continue to be computerized.

Unlike jobs in the private sector, job openings in government are usually not advertised, so you will have to do some research to find them. Information about jobs with the federal government may be obtained from any Federal Job Information Center, listed in the telephone book under U.S. Government offices. Contact your state's nearest employment services office or your library for the latest listing of job openings and how to apply. You will have to take the civil service examination before you can be considered.

# BANKING, INSURANCE, AND REAL ESTATE

BANKING, INSURANCE, AND REAL ESTATE FIRMS hire many data processing professionals.

Banks are competing in developing new services. Many now have ATMs (automated teller machines) and are forming national networks, linking their ATMs so that customers can have ready access to funds not only in their own region but across the country. They are also experimenting with debit cards through which the cost of a purchase will be automatically deducted from the customer's bank account. Additional jobs will be generated by automated check clearing facilities, electronic funds transfer systems, and centralized credit checking and authorization systems.

Local insurance agents are purchasing software to link their minicomputers with the home office's system. Systems must be developed to integrate these data.

Large real estate consortiums need systems that will accept information concerning property for sale or wanted from individual agents and disburse it to member agents.

In general, programmers, systems analysts, and technicians will be needed to develop new systems that will consolidate the services and functions of banking, insurance, and real estate.

## WHOLESALE AND RETAIL TRADE

WHOLESALE AND RETAIL TRADE FIRMS are expected to increase their employment of computer professionals by about 70 percent in the 1980s. The number of systems analysts, programmers, and computer and peripheral equipment operators will all increase, but the greatest need will be for computer service technicians.

Large mail order retailers, like Sears and Montgomery Ward, will be expanding their systems as the trend toward mail order buying continues. Shop-at-home services by which merchandise is displayed over cable television channels will become more available to the public, and the necessary systems will be developed. The Universal Product Code (the series of parallel black lines you find on nearly every product) has opened the door to further computer applications. Some grocery and department stores are now experimenting with systems that will reorder stock automatically when the computer sees that the store is running low. The warehouse computer coordinates all deliveries for efficiency and economy. And eventually systems will be created to link warehouse computers with the wholesalers' computers so that reorders are placed and filled automatically. Computer personnel will be needed to implement and link these new systems.

*Major Employers*

**SERVICE COMPANIES**
Automatic Data Processing, Clifton, NJ
Burroughs Corporation, Detroit, MI
Computer Sciences Corporation, El Segundo, CA
Electronic Data Systems Corporation, Dallas, TX
Tymshare, Inc., Cupertino, CA

**MANUFACTURING FIRMS**
Auto-Trol Technology Corporation, Denver, CO
Conrac Corporation, Stamford, CT
General Electric Company, Schenectady, NY
Gerber Scientific, Inc., South Windsor, CT
Westinghouse Electric Corporation, Pittsburgh, PA

**BANKING, INSURANCE, AND REAL ESTATE FIRMS**
Aetna Life and Casualty, Hartford, CT
Chase Manhattan Bank, New York, NY
Citibank, New York, NY
GEICO, Washington, DC
Provident Life and Accident Insurance Company,
    Chattanooga, TN

**WHOLESALE AND RETAIL TRADE**
Farm House Foods, Milwaukee, WI
K-Mart, Troy, MI
J. C. Penney Company, New York, NY
Sears, Roebuck and Company, Chicago, IL
Wal-Mart Stores, Inc., Bentonville, AK

## Career Paths

| LEVEL | JOB TITLE | EXPERIENCE NEEDED |
|---|---|---|
| Entry | Programmer trainee | College degree |
| 2 | Programmer analyst | 1 year |
| 3 | Senior programmer or senior analyst | 2 years |
| 4 | Lead programmer or lead analyst | 3-5 years |

## Job Responsibilities

### Entry Level

**THE BASICS:** Working as part of a team to design and write programs, debug programs, write documentation for programs, and devise and maintain data processing systems.

**MORE CHALLENGING DUTIES:** Working on more difficult projects. Heading up one or two simple projects.

### Moving Up

Doing good work is the first and foremost prerequisite for moving up. If you complete your projects on time and complete them well, you should have no trouble moving into a senior or lead programming job. The more you understand the work of the firm or industry in which you are employed, the better you will be able to serve its needs with efficient programs that consistently get the job done.

If you want to move into a companywide managerial position, it is easier to do so from a job in systems analysis, for there you are already dealing with others in the company or with decision-makers in client firms. You will have to demonstrate not only technical ability, but also the ability to manage a staff, to control the flow of work, and to make decisions that improve the productivity of the firm and the client.

## Salaries

| | |
|---|---|
| Trainee | $12,000 to $15,000 |
| Programmer/analyst | $15,000 to $22,000 |
| Senior programmer/senior analyst | $18,000 to $28,000 |
| Lead programmer/lead analyst | $21,000 to $36,000 |

Salaries will vary from industry to industry; they will also depend on the location and size of the firm that employs you. Keep in mind that the same job title can mean different things to different companies.

## Working Conditions

*Hours:*   In general you will keep regular business hours, although in some jobs, especially those with government or service companies, night and weekend work may be required, especially during rush periods.

*Environment:*   Entry-level personnel can expect to work in a bullpen situation, sharing a cubicle and a phone with one, two, or three other people. Some firms provide separate cubicles, but this is an exception to the rule.

*Workstyle:*   Most of the day is spent in front of the computer terminal, the rest of it in meetings. Work will be more pressured when there are deadlines to be met.

*Travel:* There are virtually no possibilities for travel at the entry level. Systems analysts with service firms may make some local calls on clients.

## SOFTWARE

The demand for new specialized software is soaring now that so many people have purchased computers for their homes and businesses. Children crave new games all the time, and parents and teachers want these to be educational as well as entertaining. Small businesses need data base and accounting programs that fit the particular needs of their industries.

Novelists and journalists want word processing programs that are easy to use, while technical writers and academics need programs that will do specialized footnoting and documentation.

Some medium-size businesses hire turnkey suppliers to set up their offices, install the best software for the job, and train their staff. Other businesses turn to service companies that design software that will transmit data over telephone lines to the service company mainframes where the processing is done for an hourly fee.

Most programs take months to create and frequently longer to debug, yet with the overwhelming demand for the new software, computer professionals will have as much work as they can handle. New developments in prepackaged software that can be easily adapted by the end-user will cut into the market for customized software, but the retail demand will take up that slack, and software publishers will be hiring more programmers and programmer/analysts than ever before.

### Job Outlook

*Job Opportunities Will Grow:* Faster than average

*Competition for Jobs:* Some

*New Job Opportunities:* End-user companies will always have a greater demand for computer professionals than the computer industry itself. Also, these are considered premium positions, more creative than the average programming jobs, so competition is stiffer.

You may find it difficult to land an entry-level position with a standard software publisher or a custom software builder, but your chances should be better with turnkey suppliers and service companies.

## How to Break into the Field

Software publishers and custom software builders look for bright people who have some background in business, accounting, and engineering. If you know what kinds of programs you would like to write, it would be in your best interests to take some courses that would help you understand the needs of that area. Also keep in mind that software publishers are clustered in certain parts of the country, so relocation may be necessary.

Large service companies and turnkey suppliers hire hundreds of programmers. With initiative and talent you can quickly move up the ranks to the divisions that do work on the most interesting projects. You will also get to see the entire process of designing and installing a system for an end-user, an exposure that will help you later on in your career.

No matter what type of company you're applying to, it is important that you show initiative and an eagerness to create new software. Do some research on the companies before you contact them. Find out what kind of software they specialize in. If you have some work experience that even remotely relates to their specialty, make sure you emphasize it in your résumé and the cover letter you send.

## Geographic Job Index

The well-known mecca of the computer industry is the Silicon Valley in northern California, and many software suppliers are clustered there. But you may not be familiar with the other centers of the computer industry: the Silicon Mountain (Colorado Springs, CO), the Silicon Prairie (North Dallas, TX), the Silicon Hills (Austin, TX), the Silicon Desert (Phoenix and Tucson, AZ), the Route 128 area circling Boston, MA, Research Triangle Park near Raleigh, NC, and the Atlanta, GA, area. Most of the software suppliers are found in these areas.

Service companies and the turnkey divisions of the large computer manufacturers are located all over the country, usually close to the larger cities where their customers are concentrated.

## Who the Employers Are

STANDARD SOFTWARE SUPPLIERS in the United States number more than 5000. Some of these are giant corporations: IBM, Tandy, Commodore, Apple, Atari, and Texas Instruments. In the next rank are companies like Lotus, Microsoft, and Intel. The remaining 28 percent of total sales in 1983 were made by the thousands of small software companies struggling to establish their positions in the marketplace. All of these companies employ programmers and programmer/analysts to create new software, but the total number of new jobs in the software industry is still only a fraction of the number in the data processing and information systems. This is one of the most competitive areas for job applicants.

Standard software suppliers create packaged software designed for general use: games, word processing, accounting, personal data base programs, spread sheets, and so forth. Some suppliers also create more sophisticated business software that can be easily programmed by people with elementary computer skills and adapted to fit specific needs. Some packaged software can be put together like building blocks to perform complicated tasks, eliminating the need for more expensive software.

Entry-level positions with software suppliers may be hard to come by because experienced programmers are usually preferred. However, if you have expertise in a certain field or course work in accounting, business, engineering, or a natural science, you will find such knowledge highly valued. Experience with a wide variety of computer languages is also a plus—specialized ones like AA as well as the more standard ones like COBOL.

**CUSTOM SOFTWARE BUILDERS** employ both programmers and systems analysts to create custom-designed software. Few firms specialize in this; most often it is just one of the services offered by the large computer service companies.

When the software is installed, product support personnel make sure that it runs properly.

Entry-level positions are available in the custom software divisions of the larger companies. Because most custom software is created for business applications, employers look for candidates who have some course work in business and accounting and prefer double majors like computer science/business administration. This is true for product support personnel as well as for programmers and systems analysts. There is also some demand for people who have formal backgrounds in engineering or the natural sciences.

**TURNKEY CONTRACTORS** are hired by companies that wish to computerize their operations. The contractor purchases and tests the hardware and software and sets up the equipment on the company's premises. In some cases, the turnkey contractor must have custom software designed for the customer. This is done by either an in-house staff of programmers and analysts or by a custom software builder. Most turnkey suppliers are divisions of the large computer manufacturers who work mainly with their own products. Entry-level opportunities in systems and applications programming do exist for recent grads in these turnkey divisions.

**SERVICE COMPANIES** provide a wide range of computer services to businesses and institutions, among them customized software. These companies mainly create software for manufacturers, government, financial institutions, insurance companies, real estate firms, wholesalers, and retailers. Both systems and applications programmers are hired to work on customized software, and double majors are preferred by far. However, you can also break in through an initial programming job in data processing. Additional course work and on-the-job experience will put you in an excellent position for moving over to the teams that create customized software.

*Major Employers*

**STANDARD SOFTWARE SUPPLIERS**
  Apple Computer, Inc., Cupertino, CA
  Commodore International, Ltd., Norristown, PA
  Informatics, Inc., Canoga Park, CA
  Mathematica Products Groups, Inc., Princeton, NJ
  Texas Instruments, Inc., Dallas, TX

**CUSTOM SOFTWARE BUILDERS**
  Control Data Corporation, Bloomington, MN
  Digital Equipment Corporation, Maynard, MA
  Honeywell, Inc., Minneapolis, MN
  ITT Corporation, New York, NY
  Management Assistance, Inc., New York, NY

**SERVICE COMPANIES**
  Automatic Data Processing, Inc., Clifton, NJ
  Hewlett-Packard Company, Palo Alto, CA
  IBM, Armonk, NY
  NCR Corporation, Dayton, OH
  TRW, Inc., Cleveland, OH

**TURNKEY CONTRACTORS**
  Digital Equipment Corporation, Maynard, MA
  Honeywell, Inc., Minneapolis, MN
  Texas Instruments, Inc., Dallas, TX
  Xerox Corporation, Stamford, CT

## Career Paths

| LEVEL | JOB TITLE | EXPERIENCE NEEDED |
|-------|-----------|-------------------|
| Entry | Programmer trainee | College degree |
| 2 | Programmer or analyst | 6 months to 1 year |
| 3 | Senior programmer or senior analyst | 2-3 years |
| 4 | Lead programmer or lead analyst | 4-5 years |

## Job Responsibilities

### Entry Level

**THE BASICS:**   Working as a member of a team to write and debug software programs.

**MORE CHALLENGING DUTIES:**   Meeting with end-users to determine their needs. Working independently or initiating projects.

### Moving Up

Moving up will mean working on more difficult and complex projects and supervising the work of others. You may choose to move out of programming to product support positions, dealing effectively with end-users, who are often computer "illiterate"; you will need excellent communications skills. Or you may want to get into the lucrative area of sales. In turnkey contracting the efficient coordination of work is a paramount concern; those with managerial aptitude—good organizational skills and the ability to manage a professional staff—can set their sights on moving into supervisory or corporate management positions.

## Salaries

Compensation varies widely according to the size of the firm and its location. As with data processing and information systems jobs, titles mean different things to different companies.

| | |
|---|---|
| Programmer trainee | $16,000 to $18,000 |
| Programmer or analyst | $18,000 to $27,000 |
| Senior programmer or senior analyst | $22,000 to $34,000 |
| Lead programmer or lead analyst | $27,000 to $35,000 |

## Working Conditions

*Hours:* Office hours are generally nine to five, but you will be expected to work as long as it takes to get the job done.

*Environment:* Bullpen-style shared office space is the norm, with two to four programmers or analysts sharing a cubicle and a telephone.

*Workstyle:* Most of the day is spent in front of the terminal; some time is spent in meetings with other programmers or with end-users.

*Travel:* At the entry level, opportunities for travel are limited. Analysts will meet with customers at their places of business. In general this is local travel.

# HARDWARE

It is predicted that in 1984 micro sales will surpass mainframe sales for the first time, reflecting the trend away from large computers and toward home and desktop computers. Companies are finding that they don't always need the computing power of a mainframe, that often a series of linked personal computers will more than adequately serve their needs. Business people are buying personal computers for themselves that are compatible with their office systems so that they can work at home. More children are clamoring for home computers than ever before, and the first generation of computer kids is now starting to upgrade their systems, buying better hardware. The victory of the microcomputer is particularly remarkable considering that eight years ago they weren't even on the market.

The phenomenal success of the micro does not mean the end of the mainframe, however. Even though sophisticated minicomputers are replacing midsize mainframes, advanced mainframes—the so-called supercomputers made by companies like Amdahl Corporation and Cray Research—are going strong. And IBM, which builds four out of every five mainframes sold, will continue to dominate the market, showing their greatest profits in this area. Some companies will suffer as a result of the invasion of the micros, though, most notably the group known as "the BUNCH" (Burroughs Corporations, Sperry's old Univac division, the NCR Corporation, Control Data Corporation, and Honeywell, Inc.). In order to stay competitive, Honeywell and Sperry are marketing Japanese-made mainframes under their names. Naturally, with Honeywell selling computers made by Nippon Electric Company and Sperry selling the Mitsubishi product, many computer hardware professionals have lost their jobs. The majority of these employees have been absorbed by the microcomputer makers.

Although IBM is expected to control the mainframe market, the microcomputer industry is a wide-open field, with many companies scrambling for a portion of the market and most of them predicted to fold in the great shakeout that will ensue. Among the microcomputer manufacturers, Apple Computer and Wang Laboratories are the strongest at the moment, but who will dominate the market in the next few years is anyone's guess. What is certain is that the companies that do survive the shakeout will emerge stronger and more profitable as a result.

## Job Outlook

*Job Opportunities Will Grow:*   About as fast as average

*Competition for Jobs:*   Keen

*New Job Opportunities:*   Computer-aided manufacture (CAM) systems are one of the most rapidly expanding areas in the industry. These systems adapt equipment and robots to perform specific factory operations. Now, instead of a machine or group of machines producing a single product, a variety of customized products with computerized modifications can be made without additional capital investment. The field of artificial intelligence (AI) is also picking up steam. This is a sophisticated discipline that attempts to teach computers to "think," so that they can almost run by themselves. Some companies are producing super mainframes, lightning-fast computers that can process and distribute vast quantities of information very quickly. As these three areas develop, they will provide new jobs in the hardware area.

## How to Break into the Field

Most jobs in the computer hardware industry are highly technical positions that require advanced engineering degrees. Programmers are hired to be part of the design teams, but these positions are not easy to come by. However, computer professionals are needed to demonstrate new products, answer technical questions about them, train customers in their use, and generally support the product once it's installed on the customer's premises. Marketing professionals have to be able to relate to people, and good communications skills are a must, since marketing representatives are the customer's chief liaison with the technology. Sales and marketing people are the best paid in the computer industry, and many programmers and systems analysts try to move into this area.

## Geographic Job Index

The major centers of the computer hardware industry are northern California's Silicon Valley; the Silicon Mountain (Colorado Springs, CO); the Silicon Prairie (North Dallas, TX); the Silicon Hills (Austin, TX), the Silicon Desert (Phoenix and Tucson, AZ), the Route 128 area near Boston, MA; and North Carolina's Research Triangle Park.

## The Work

Few computer manufacturers actually manufacture all the components that make up the computers they market. They commonly contract with other manufacturers for components and assemble them to make their computers. Computer personnel in hardware firms usually work as part of the design team, the prototype team, or the manufacturing team.

The design team is made up chiefly of electrical engineers, systems engineers, mechanical engineers, and software specialists. The project leaders on the design team are almost always engineers, but programmers are becoming more integral parts of these teams with the recent introduction of firmware—software that is permanently programmed into the computer's ROM (read only memory).

The prototype team actually builds the computer and tests it for bugs, or problems. The team consists of electronics specialists, assemblers, inventory control specialists, and materials specialists.

When the perfect prototype has been built, the manufacturing team takes over and devises the optimum production method. Industrial, electrical, and mechanical engineers are the nucleus of the manufacturing team.

Hardware sales and marketing require a high degree of technical knowledge; sales personnel have to be able to compare their product with the competition's and pinpoint the advantages of one computer over another. The sales and marketing department is made up of marketing engineers, field sales representatives, systems representatives, and product support representatives.

*Major Employers*

Amdahl Corporation, Sunnyvale, CA
Data General Corporation, Westboro, MA
Datapoint Corporation, San Antonio, TX
IBM General Products Division, Tucson, AZ
Wang Laboratories, Lowell, MA

## Career Paths

| LEVEL | JOB TITLE | EXPERIENCE NEEDED |
| --- | --- | --- |
| Entry | Programmer | College degree, previous programming experience |
| 2 | Senior programmer | 2-4 years |
| 3 | Lead programmer | 4-5 years |

## Job Responsibilities

### Entry Level

THE BASICS: Working as a member of a team on simpler elements of hardware design.
MORE CHALLENGING DUTIES: Tackling more difficult components of a job. Team leadership.

### Moving Up

Programmers in the hardware field generally remain part of the design team, moving on to solving the more elusive problems that arise in the course of designing hardware. They rarely move up into management. Management positions most often go to engineers. In sales and marketing, however, people are not restricted by their educational backgrounds. A solid track record is the major criterion for promotions into management.

## Salaries

**PROGRAMMING**

| | |
|---|---|
| Programmer | $16,500 to $26,900 |
| Senior programmer | $19,200 to $33,100 |
| Lead programmer | $23,800 to $38,200 |

**MARKETING**

| | |
|---|---|
| Product support representative (1 to 4 years' experience) | $16,800 to $35,400 |
| Product support representative (4+ years' experience) | $26,900 to $41,500 |
| Marketing representative (salary plus commissions) | $26,000 to $80,000+ |

## Working Conditions

*Hours:*   These are basically nine-to-five jobs, with overtime required when deadlines must be met.

*Environment:*   Expect to share a cubicle and telephone with other programmers.

*Workstyle:*   Work is done at your terminal or in meetings with other members of your team.

*Travel:*   At senior levels there may be some travel to conferences or other company divisions, but virtually none for entry-level programmers. Marketing and product support representatives have wider opportunities for travel to clients' place of business.

# TRAINING AND TEACHING

Some experts predict that computer training and teaching will be the biggest growth area of the future. The trend toward teaching computer courses in high schools and even elementary schools will continue as microcomputers are used more routinely in education. Right now some of the more interesting computer teaching positions can be found at computer camps, where kids are taught computer science along with swimming, camping, and horseback riding. The Original Computer Camp in Santa Barbara, CA, and the New England Computer Camp in Moodus, CT, are two of the better known ones.

In the next few years there will be an even greater number of jobs teaching computer courses to adults. Office personnel must be trained to use the new equipment of the automated office: word processing programs, spread sheets, and data base packages. There are companies that specialize in this kind of training; many more are expected to appear in the next few years. Commonly, sales personnel from these computer training firms assess the customer company's needs, taking into consideration what equipment they have, what they are planning to buy, and what software they want their personnel to learn. Courses are then planned, and instructors are assigned.

Training firms do not deal with office personnel exclusively. As new technology enters the marketplace, computer professionals must be retrained, too. Recent developments in new operating systems and advanced computer languages have created a demand for instructors who can train experienced computer personnel.

The traditional educational book publishers are now in the software business, employing computer-savvy editors and programmers to create CAI (computer-aided instruction) packages for elementary, junior high and high-school subjects. People will be needed to review and edit books and manuals concerning all aspects of computers, and these computer editors will be in particular demand for CAI projects. The major educational publishers working in this area now are Harcourt Brace Jovanovich, Inc., New York, NY, Macmillan Publishing Co., New York, NY, Houghton Mifflin Company, Boston, MA, Harper & Row, New York, NY, and Ginn and Company, Lexington, MA.

As computers become increasingly user-friendly, the majority of the new job opportunities will be in people-oriented areas. The number of instructors who can show people how to use computers effectively and teachers who can create and use CAI packages may overtake the number of programmers and systems analysts by the 1990s.

## ENTREPRENEURS

Steve Jobs and Steve Wozniak, who developed Apple Computers in a California garage, exemplify an entrepreneurial spirit that has been the driving force behind the computer industry. The industry itself is less than 20 years old, yet phenomenal changes have taken place. Desktop microcomputers didn't even exist eight years ago; today they have overtaken the industry.

Startup companies appear every week, but entrepreneurs need not set up their own companies to be successful. Individuals who usually call themselves consultants have made some important contributions to the industry.

Although they seem to defy generalizations, one can say that all computer entreprenuers fill a need, whether it's in creating software for new applications, building specialized equipment, writing custom programs, consulting, retailing, teaching, or publishing computer magazines. Without a doubt, they all share foresight, imagination, the willingness to take risks, and the desire to succeed.

# INTERVIEWS

**Hugh Blair, Age 35**
**Technical Support Manager**
**Blue Sky Software, Inc., Cherry Hill, NJ**

I've always been fascinated with electronics. When I was in high school, I built the first digital clock my town ever saw. It cost something like $400 to make because the chips that sell for about ten cents now went for something like $4 then. I didn't go on to college to major in computer science or electrical engineering, though. Instead I took two years off between high school and college and worked as an assistant manager at a hotel in Colorado. When I finally got to college, the thought of working with circuit boards and wires didn't appeal to me anymore, and I wound up majoring in hotel management.

From hotel management, I eventually went on to manage a print shop, which is where I first started to use a computer. I like to figure out how things work, so it didn't take long before I had taught myself how to program. Unfortunately, the print shop was sold out from under me, and suddenly I was unemployed—one of the lowest points in my life. That period lasted nine months, but in that time I got my hands on a Commodore and became addicted to it. It wasn't unusual for me to spend 10, 12 hours a day with the computer. I just tried to do everything imaginable with it. A friend of mine subscribed to a lot of computer magazines, and after he was through with them, I devoured them. I absorbed as much as I possibly could about home computers.

One day I read an ad in the help-wanted section of the paper. An exclusive girls' camp in Maine was looking for a computer operator to do their scheduling. The camp had 200 girls, 40 different activities, and five periods in their day, but they always seemed to have too many kids in the lake when no one was on the tennis court, problems like that. The director of the camp thought he could solve all his scheduling problems with a computer, so he went to Radio Shack and bought $5000 worth of equipment—except they didn't sell him any software. He was tearing his hair out because he couldn't get the computer to do what he wanted. When I got the job, I wrote a program for scheduling, and the director was so happy he not only paid me handsomely with $10 a day extra for traveling, he gave me a $1500 bonus at the end of the summer.

From that job, I decided to go into business for myself. In 1979, there was only one computer store in the whole state of Maine. I decided they could use some competition, so I called Commodore and told them I wanted to be a dealer. They said great, and within a year we had an annual volume of $3 million. One year we sold more Commodores to educational institutions in Maine than Apple and Radio Shack combined.

After I sold that business, I moved on to Blue Sky Software, Inc., a company from which I'd bought software for the store. My position here at Blue Sky is technical support manager, which means I'm the guy who answers the customer's questions when he says, "But I can't get the screen to turn purple!" I also represent the company at shows and conventions, and I bring in new software, evaluating and preparing it for the marketing department. I've bought some great programs from hackers as far away as Japan and Australia, as well as from local people here in New Jersey. We sell everything in software from games to the latest technology in program generators—software that lets you talk to it in English while it writes the program.

Microcomputers are the future, but I'm amazed to see how little recent grads know about micros. College computer science programs concentrate on minis and mainframes; students learn about bits and bytes and accounting applications. Now that's fine if you plan to get into data processing in banking or insurance or something like that, but it's just not applicable to the micro business. Give me an eighteen-year-old who's spent one year working on a home computer and I'll hire him on the spot. He'll have my job in six months, and I'll be able to move up.

One big problem computer science graduates have these days is the confusion of job titles in the computer industry. Job responsibilities are not as cut and dried as people have been led to believe. A programmer at one company may do something very different from his counterpart at another company. And as far as systems analysts are concerned, they don't really exist in the micro industry. A good dealer will be able to tell you what you need. Or better yet, join a users' group and ask your fellow members what you should buy in terms of hardware and software. The consensus of opinion in a users' group is going to be worth a lot more than the advice you'd get from a single systems analyst.

My advice to someone who wants to get a job in the microcomputer industry? Be a sponge. Absorb as much information as you can. Buy yourself a home computer, the best you can afford, then run every program you can get your hands on. Experiment with it, learn it inside out. Read every computer magazine you can find. Join a users' group and pick the members' brains, share information. I'm always picking up little nuggets of information from people in my users' group. Once you've really immersed yourself in home computers this way, then you're a hot property as far as I'm concerned. And that's the kind of person I like to hire.

**Neil Hunt, age 26**
**Product Support Specialist**
**Avco Computer Services, North Andover, MA**

I graduated from North Adams State College in Massachusetts in June 1979 with a major in business administration and a minor in computer science. I studied COBOL I and II, Assembler, MIS, and several theoretical system design languages, but I consider the time I spent on the job with Sprague Electric through the college's cooperative program just as valuable as my course-work.

The recent graduates I've met know a lot about computer theory, but their courses usually don't show them how to apply it to business. Knowing advanced computer languages is great, but if you can't comprehend an accountant's problem that can be solved using COBOL, you're not ready to walk into a programmer's job in the real world.

I started as a programmer with Bradford Trust in Boston, using COBOL to write programs, as well as testing and debugging programs. From there, I moved on to Star Market, a supermarket chain, writing programs for payroll, benefits, and accounting. Eventually I worked my way up to project leader with two programmers under me.

Here at Avco I'm a product support specialist, training and consulting with end-users on our payroll/human resources software package. You can say I'm the middleman between the hardware vendor and the end-user, making sure the system does everything it was designed to do. Though I don't write programs in this position, my experience as a programmer was instrumental in my getting to this level.

Eventually, I'd like to get into sales of software packages. I'm in an M.B.A. program right now, so obviously I decided not to pursue a technical track. At this time, a lot of people can write programs, and that's very important work, but I think those who can write programs and talk about them intelligently are really special. That's an asset I have and I'm proud of it. I feel that sales is the natural direction for me because it combines my best qualities: my ability to understand how a system or a program works and my ability to explain it to a potential customer.

# DEPARTMENT STORE RETAILING

CONSUMERS generally take for granted that they will always find their favorite department stores brimming with merchandise. Unnoticed by most customers, a large, talented staff works long, hard hours to keep the shelves filled, the selection varied, the stores beautiful, and the business of retailing running smoothly. Retailing is an industry in which brains and diligence can take you to high levels of decision-making years before your contemporaries in other fields have reached similar positions of responsibility.

Graduates of virtually any discipline may enter department store retailing. Prospective employers are looking for demonstrated capacity to learn and make quick, sound judgments and are less interested in academic backgrounds. You must be flexible, comfortable with people, self-disciplined, and highly motivated—and a sense of humor certainly does not hurt. Retailing is a high-pressure profession where no slow seasons exist—only busy and busier, with the November-December pre-Christmas rush being the most hectic time of all. Prior retail experience, even a summer spent behind a cash register, is a plus; some retailers won't consider candidates without it.

Most entry-level jobs are in merchandising, an area further divided into:

- **Store Management**
- **Buying**

Your job in merchandising begins with a training period of six months to a year. Some trainees divide their time between classroom learning and work experience, others train entirely on the job. Generally, the larger the retailer, the more formalized the training. Whether you enter the field via store management or buying depends primarily on the employer. Many stores separate these functions beginning at the entry level; you must choose which path you prefer. Other stores will introduce all new merchandising personnel to buying and later allow those interested in and qualified for management to move up. The opposite arrangement, moving into buying at some later stage, also occurs, although infrequently.

The modern store is reaping the benefits of the technological revolution. Point-of-sale computer terminals are replacing mechanical cash registers; these automatically compute sales, taxes, and discounts and simplify inventory control by keeping sales records. Computers are also used for credit records and tracking sales forecasts.

Retailing is vulnerable to downturns in the economy, but it's one of the first industries to bounce back after a recession. As a highly profit-oriented business, it's hectic and competitive. The customer's satisfaction and loyalty to the store are very important, which means that you must tolerate and even pamper people whom you may not like. In retailing, the unexpected is the order of the day; you can expect to feel pressured, but seldom unchallenged.

## Job Outlook

*Job Openings Will Grow:*    As fast as average

*Competition for Jobs:*   Keen
In merchandising, the most competition exists in buying; this area has fewer openings, tends to pay a bit better, and has an aura of glamour about it.

*New Job Opportunities:*   An exciting new technological development, still in experimental form, that may change retailing in the next decade is video retailing. A select number of communities now have a two-way cable television system through which viewers may receive and send information to a broadcasting center. Viewers can order goods seen on the screen by typing their selections on a keyboard. Video retailing is still in developmental form, but those entering retailing should be aware of its potential as a new job area.

## Geographic Job Index

The location of retail jobs parallels the distribution of the general population; stores operate where customers live. As an up-and-coming executive in a retail chain, expect to work in a city or suburban area. Most new store construction in the coming years is expected to take place in revitalizing city cores. Department stores are found across the country, with the highest concentration of jobs in the Northeast, Midwest and West Coast.

If your interest is buying, your geographic options are more limited. For many department store chains, most or all buying takes place in a few key markets, notably New York, NY.

## Who the Employers Are

A retailer is, in its simplest definition, a third party who sells a producer's goods to a consumer for a profit. The retailing industry as a whole comprises a wide variety of stores of different sizes with different personnel needs. Management personnel are sought by all major retail firms, including grocery, drug, specialty, and variety store chains, but because the most varied opportunities are found in department stores, this chapter focuses on this sector of retailing.

*Major Employers*

Allied Stores Corporation, New York, NY
Bonwit Teller
Field's
Jordan Marsh
Stern's

Carter Hawley Hale Stores, Los Angeles, CA
Bergdorf Goodman
The Broadway
John Wanamaker
Neiman-Marcus

Dayton Hudson Corporation, Minneapolis, MN
Dayton's
Diamond's

Federated Department Stores, New York, NY
Abraham & Straus
Bullock's
Filene's
Foley's
I. Magnin
Rich's

R.H. Macy & Company, New York, NY

Montgomery Ward & Company, Chicago, IL

J. C. Penney Company, New York, NY

Sears, Roebuck & Company, Chicago, IL

### How to Break into the Field

Your best bet is on-campus interviews. Major retailers actively recruit on college campuses. This is the most accessible way to most potential employers. Don't hesitate, however, to contact employers directly, especially if you want to work for a smaller operation. Read the business section of your newspaper regularly to find out about store expansions, the addition of new stores or locations, and other developments in retailing that can provide important clues to new job openings. Keep in mind that retail or selling experience of any kind will increase your chances of getting hired.

### International Job Opportunities

Extremely limited. Opportunities to live abroad exist at the corporate level of a few international chains.

## STORE MANAGEMENT

If you're a "people person," consider the store management side of merchandising. You'll be responsible for handling the needs of staff and customers.

The job of store management personnel, even at entry level, entails making decisions on your own. But since decisions often have to be made on the spot and involve balancing the interests of both customers and the store, your mistakes are likely to be highly visible. Whether you manage the smallest department or a very large store, you must always keep the bottom line—making a profit—in mind when making decisions.

During training, you will work with experienced managers and will be moved throughout the store to observe all aspects of merchandising. If you're quick to learn and demonstrate management potential, you'll soon be made manager of a small department or assistant manager of a large one. You will have a fair amount of autonomy, but you must stick to store standards and implement policies determined by higher level management.

## Qualifications

*Personal:*   Ability to learn quickly. Enormous enthusiasm. The flexibility to handle a constantly changing schedule. Willingness to work weekends, holidays, and nights.

*Professional:*   Demonstrated leadership ability. Ability to work with figures, finances, inventories, and quotas. A sense of diplomacy.

## Career Paths

| LEVEL | JOB TITLE | EXPERIENCE NEEDED |
|-------|-----------|-------------------|
| Entry | Department manager trainee | College degree |
| 2 | Group department manager | 2-3 years |
| 3 | Assistant store manager | 5-10 years |
| 4 | Store manager | 8-12 years |

## Job Responsibilities

Entry Level

THE BASICS:   Handling staff scheduling. Dealing with customer complaints. Doing plenty of paperwork.

MORE CHALLENGING DUTIES:   Monitoring and motivating your sales staff. Assisting in the selection of merchandise for your department. Making decisions and solving problems.

## Moving Up

Advancement in store management depends on how well you shoulder responsibility and take advantage of opportunities to learn. Effectively leading your staff, moving merchandise, and, above all, turning a profit will secure your promotion into higher levels.

Your first management position will be overseeing a small department, handling greater volumes of money and merchandise. The group department manager directs several department managers, coordinating store operations on a larger scale. From here you might progress to assistant store manager and store manager; this last position is, in many respects, similar to running a private business. The best may then go on to the corporate level.

Relocation is often necessary in order to win promotions. Switching store locations every three years or so is not uncommon. However, depending on the chain, a change of workplace need not require a change of address; often stores are within easy driving distance of each other. But the larger the chain, the greater the possibility that you'll have to move to a different city to further your career.

# BUYING

Do you fantasize about a shopping spree in the world's fashion capitals? A few lucky buyers, after years of work and experience, are paid to do just that when they're sent to Hong Kong, Paris, or Milan to select new lines of merchandise. Most do not make it to such heights, but on a smaller scale, this is the business of buying.

A buyer decides which goods will be available in a store. Buyers authorize merchandise purchases from wholesalers and set the retail prices. A sensitivity to changing trends, tastes, and styles and an ability to understand and forecast the preference of your own store's customers is crucial. Buyers must also maintain standards of quality while keeping within certain ranges of affordability.

The buyer who works for a discount department store faces a particularly tough job. Obtaining lower-than-average prices for quality merchandise is a real challenge and requires an unerring eye and an ability to negotiate with sellers.

Astute buying translates into profits for the store and advancement for your career. Learning how to spend large sums of money wisely takes practice. Fortunately, as a new buyer you can afford to make a few mistakes, even an occasional expensive one, without jeopardizing your career. A good buyer takes calculated risks, and as you gain experience more of your choices will succeed.

During training, you'll work immediately as an assistant to an experienced buyer. The trainee progresses by observing, asking questions, and offering to take on appropriate responsibilities.

## Qualifications

*Personal:*    An interest in changing trends and fashions. An ability to work with a wide variety of personalities. A willingness to channel creativity into a commercial enterprise.

*Professional:*    Financial and negotiating know-how. Organizational skills. Good judgment in spotting trends and evaluating products.

## Career Paths

| LEVEL | JOB TITLE | EXPERIENCE NEEDED |
|---|---|---|
| Entry | Assistant or junior buyer | College degree and store training |
| 2 | Buyer (small lines) | 2-5 years |
| 3 | Buyer (large lines) | 4-10 years |
| 4 | Corporate merchandise manager | 15+ years |

## Job Responsibilities

## Entry Level

THE BASICS: Assisting your supervising buyer. Placing orders and speaking with manufacturers by phone. Supervising the inspection and unpacking of new merchandise and overseeing its distribution.

MORE CHALLENGING DUTIES: Becoming acquainted with various manufacturers' lines. Considering products for purchase. Evaluating your store's needs. Keeping an eye on the competition.

## Moving Up

Advancement depends on proof of your ability to judge customer needs and to choose saleable goods. The only purchases closely scrutinized by higher authorities are those inconsistent with past practices and standards.

After completing your training, you will first buy for a small department, then, as you become seasoned, for larger departments. High-placed buyers make decisions in buying for a key department common to several stores, for an entire state, or possibly for many stores. Your buying plans must always be well coordinated with the needs of store management.

# ADDITIONAL INFORMATION

## Salaries

Entry-level salaries range from $12,000 to $18,000 a year, depending on the employer and the geographic location of the store. Junior buyers tend to be among the best paid entry-level employees.

The following salary ranges show typical annual salaries for experienced retail personnel. In merchandising salaries vary with the size and importance of your department.

| | |
|---|---|
| 2-4 years: | $16,000-24,000 |
| 5-10 years: | $22,000-27,000 |
| 12 years or more: | $25,000 and up |

## Working Conditions

*Hours:*   Most retail personnel work a five-day, 40-hour week, but schedules vary with different positions. In store management, daily shifts are rarely nine to five, because stores are open as many as 12 hours a day, seven days a week. Night, weekend, and holiday duty are unavoidable, especially for newcomers. Operations personnel work similar hours. Buyers have more regular schedules and are rarely asked to work evening and weekend hours.

*Environment:*   In merchandising, your time is divided between the office and the sales floor—more often the latter. Office space at the entry level may or may not be private, depending on the store. Whether you share space or not, expect to be close to the sales floor. Merchandising is no place for those who need absolute privacy and quiet in order to be productive.

*Workstyle:*   In store management, office time is 100 percent work; every valuable moment must be used effectively to keep on top of the paperwork. On the floor you will be busy overseeing the arrangement of merchandise, meeting with your sales staff, and listening to customer complaints. Long hours on your feet will test your patience and endurance, but you can never let the weariness show. In buying, office time is spent with paperwork and calls to manufacturers. You might also review catalog copy and illustrations. On the sales floor, you'll meet with store personnel to see how merchandise is displayed and, most important, to see how the customers are responding. Manufacturers' representatives will visit to show their products, and you might spend some days at manufacturer and wholesaler showrooms. Because these jobs bring you into the public eye, you must be well dressed and meticulously groomed. The generous discounts that employees receive as a fringe benefit help defray the cost of maintaining a wardrobe.

*Travel:*   In store management, your responsibility lies with your own department and your own store; travel opportunities are virtually nonexistent, except for some top-level personnel. Buyers, particularly those who live outside major manufacturing centers, may make annual trips to New York, NY, and other key cities. You might also travel to trade shows at which your type of merchandise is displayed.

### Extracurricular Activities/Work Experience

Leadership in campus organizations

Treasurer or financial officer of an organization

Sales position on the yearbook or campus newspaper

Summer or part-time work in any aspect of retailing

### Internships

Arrange internships with individual stores or chains; many are eager to hire interns, preferring students who are in the fall semester of their senior year. Check with your school's placement or internship office or with the store itself in the spring for a fall internship. Summer internships are also available with some stores. Contact the placement office or the personnel departments of individual stores for details.

# Recommended Reading

**BOOKS**

*Buyer's Manual*, National Retail Merchants Association: 1979

*Creative Selling: A Programmed Approach* by R.J. Burley, Addison-Wesley: 1982

*The Retail Revolution: Market Transformation, Investment, and Labor in the Modern Department Store* by Barry Bluestone  et al., Auburn House: 1981

*The Woolworths* by James Brough, McGraw-Hill: 1982

**PERIODICALS**

*Advertising Age* (weekly), Crain Communications, 740 North Rush Street, Chicago, IL 60611

*Journal of Retailing* (quarterly), New York University, 202 Tisch Building, New York, NY 10003

*Stores* (monthly), National Retail Merchants Association, 100 West 31st Street, New York, NY 10001

*Women's Wear Daily* (daily), Fairchild Publications, Inc., 7 East 12th Street, New York, NY 10003

# Professional Associations

American Marketing Association
250 South Wacker Drive
Chicago, IL 60606

American Retail Federation
1616 H Street, N.W.
Washington, DC 20006

Association of General Merchandise Chains
1625 I Street, N.W.
Washington, DC 20006

National Retail Merchants Association
100 West 31st Street
New York, NY 10001

# INTERVIEWS

**Carolyn Egan, Age 33**
**Fashion Coordinator**
**Bloomingdale's Department Store, NY**

My first job was far removed from retailing—I taught high school math for a year. But the school environment really didn't excite me and I felt I could get more from a job. I saw an ad for the position of fashion coordinator at a branch of Gimbels' department store. I wasn't planning a career in retailing, but because I kept up with fashion and felt I had a flair for it, I applied. I got the job and enjoyed the work, but that particular branch was not a high-caliber store, and after two years I was ready to move on.

I took a part-time job as an assistant manager at an Ann Taylor store, one of a chain selling women's clothing. At that time I was also going to school to finish an art degree. My job included store management and some limited buying. I wound up managing my own store, but because Ann Taylor has a small management staff, I felt there wasn't enough growth potential. I came to know the man who was doing store design for the chain. He was expanding his operations and needed help, so I went to work with him. I designed store interiors and fixtures, which gave me a whole new perspective on the industry. I have been lucky to see so many sides of retailing, but these job changes also required me to relocate.

When I moved into fashion coordination with Bloomingdale's about seven years ago, I finally found what I had been looking for—a high-powered, high-pressured environment. When I walk into the store each morning I feel that things are moving, happening. That's the fun of retailing.

My responsibility is to work with the buyers, helping them choose the right styles. After you've been in retailing a number of years, you know where fashion has been and you can see where it's going. You decide—really by making educated guesses—what the public will want a year from today. My job includes a lot of travel—usually eight or nine weeks a year. Where there are products abroad, we explore them. That's the only way to keep up with the competition.

In buying we speak of hundreds of dozens, so you must be volume-oriented. You ask, "What does our regular customer want to see?" Then you make a decision that has to be more right than wrong. I work with children's wear, a department that rarely sees radical changes in style. But there are always new trends in color and design, and new products.

One of the toughest parts of my job is training new buyers and helping with their first big buys. They are understandably nervous about spending several hundred thousand dollars. The fashion coordinator is one with buying experience. You offer better advice if you understand the pressure and monetary responsibility of the buyer's job.

Even though I'm in a creative area, business and financial concerns are of the highest importance. You must have a head for business in every retailing job. You want to find beautiful quality products, but if they don't sell, you've failed.

The one drawback to my job is advancement. My talents and experience are best used right where I am now. Unlike the buyers, I really have no place higher to go. But I enjoy my work. I suppose it's like being an artist, and how many artists are really appreciated?

## G. G. Michelson, Age 58
## Senior Vice President for External Affairs
## R. H. Macy & Company, NY

My job is a rather unique one—it had never existed before and was tailored just for me. I represent the company in the community in its relationships with government, and in philanthropy. I was the senior vice president for personnel and labor relations in the New York division before moving into the corporate side about five years ago.

I was given this opportunity because of my long association and familiarity with the company and the business. We have a separate public relations department, and I don't interfere with their plans; rather, I am involved in considerations of corporate policy. For example, I handle difficult shareholder and community questions. We have a substantial philanthropy budget to work with. We want to spend this money creatively, but our charitable actions must be in line with our business decisions. We are primarily concerned with the communities in which our stores are located, because we recognize our obligation to those places in which we make our living.

I was quite young when I graduated from college, so I went on to law school to mature and get that valuable credential—but I never intended to practice law. All along I knew that I wanted to work in labor relations.

I considered manufacturing and some of the heavy industries as potential employers, and I came to realize that retailing as a service industry was far more people-intensive than other businesses. I found that in retailing the personnel function had a great deal more status and received more attention from top management. Looking elsewhere, I noticed that the emphasis was on cost control, not people development.

I went directly into Macy's training program from law school. The training program was and, of course, still is largely devoted to merchandising. I worked in merchandising only for the six months that I trained, but that experience gave me an excellent background for understanding the business and the people in it. In employee relations, I had responsibility for hiring, training, and developing our employees and merchandising talent.

In the past ten years, I have seen a significant change in the kind of graduates entering retailing. We now hire a great many graduates who once would have pursued other careers—graduates certified to teach, for instance—and people with liberal arts backgrounds who once would have gone on to grad school. We have always hired people who have broad educations; we have never been too concerned about a candidate's business background. We develop our talent by training people for top management, so we are looking for the ability to learn and grow. We don't want to have to train a person to think for the first time!

I spend a lot of time seeing and counseling young people who are investigating careers. My advice: be expansive and open to unforeseen opportunities. So many graduates have rigid plans— which I jokingly refer to as their "five-year plans." Often the best things that happen in a person's career development are totally unexpected. Bright people should be more flexible than many seem to be.

# EDUCATION

M ANY of the world's most prominent citizens have either started their careers as teachers or have added teaching to the list of their accomplishments. Leonardo da Vinci, Leo Tolstoy, Henry Kissinger, and Jimmy Carter are but a few.

Through the ages the theory of education has fascinated many. Tolstoy put off writing his novels in order to work out his ideas about education, and Plato devoted many of his philosophical treatises to exploring how knowledge is transmitted from teacher to student. People teach, he said, not by writing books or making speeches, but by becoming vitally involved in dialogue, in human relationships. Teaching is something that happens between two people; it is communication, knowledge passing from one to another.

Thousands of years later this still holds true. Ask almost any teacher what constitutes good teaching and he or she will tell you: good teaching is caring for the people you teach.   There are less altruistic reasons for becoming a teacher, however, and if you ask a veteran, he or she will probably start with summer vacation—not to mention the three or four weeks of holidays that can accumulate at Christmas, Easter, Thanksgiving, and the scattered celebrations of America's heroes.

Teaching can often serve personal interests, too. For those who love to act, teaching provides the perfect audience. For those who like to be in charge, teaching is one of the few professions where you start off as the boss. For scholars, teaching is a way to delve further into your favorite subjects.

Of course, teaching is not the ideal state. Although the starting salaries can be commensurate with the job market as a whole, salaries for experienced teachers most definitely are not. Crime in the schools has received a lot of national coverage, but what is called crime by the media usually means discipline problems in the school, such as absenteeism, vandalism, and abuse of drugs or alcohol. But prospective teachers must take into account the disciplinary difficulties some of them will encounter.

Perhaps the most troubling problem facing a person interested in a teaching career is the beating teachers' reputations have taken in the last decade. As the quality of education seems to have gone steadily downward, largely because of factors beyond teachers' control, teachers themselves have had to take on more and more responsibilities—not only in the area of teaching, but in the area of genuine care and concern for their students.

Few professions, however, have more social significance than teaching, and none offers the sense of satisfaction that comes from contributing directly and positively to a young person's future. When Wally Schirra, one of the original seven U.S. astronauts, was asked who was the most influential person in his life, he didn't hesitate to say his second-grade teacher.

As computers become standard equipment in more and more classrooms, teacher applicants with computer skills are in a better position to be hired. The educational value of computers, however, depends on how teachers use them. The computer can be used as an automated taskmaster or, more importantly, as an interactive device. Some programs teach sophisticated skills such as thinking and writing even to very small children. Computers can grab and keep a student's attention, stimulate, and motivate him or her to higher levels of achievement. But it is still up to the teacher to provide the crucial interaction that makes learning possible.

# Job Outlook

*Job Openings Will:*   Decline

*Competition For Jobs:*   Keen

*New Job Opportunities:*   If current migration trends continue, a 25% decline in the number of 15 to 19-year-olds is projected in the northeast and north central states between 1980 and 1990. The rate of decline is expected to be 15% in the rest of the country. These statistics, combined with the fact that the 61,650 secondary school teacher applicants from the graduating class of 1981 exceeded the number of job openings by 18,150, indicate  that new job opportunities are not plentiful. These statistics do not take into account, however, the graduates who applied for and subsequently accepted positions with independent schools.

Because of the recent budget cuts, school systems have suffered severe cutbacks in expenditures in all areas including teacher hiring and salary increases. But positions do exist, especially in math and computer sciences. Many teachers retire, leave the profession, or find administrative positions, creating thousands of openings each year. Willingness to relocate increases their chances of finding a position.

In 1978 a law was passed mandating special education for all handicapped children. The number of special education teachers rose more than 30 percent between 1970 and 1980, to 187,900. The Department of Education is still predicting a shortage of special ed teachers as the population of children needing special education continues to grow. Job supply is not the only incentive for entering this area; generally, special ed teachers are paid anywhere from $500 to $2000 more per year than other teachers.

Many parents are turning to church-affiliated institutions for the education of their children. Although salaries are almost always lower, job prospects in this area will grow.

New opportunities also exist in the areas of continuing education and preschool education. Adult education courses on virtually every subject imaginable are taught all over the country at local libraries and community centers. The need for nursery schools, some operated by teachers from their own homes, is growing as more and more mothers of young children work full- or part-time.

## Geographic Job Index

The majority of jobs are found in suburban school systems and city schools; fewer positions are available in rural and remote regions. New York, California, Texas, Pennsylvania, Michigan, and Illinois have the greatest teacher populations and the greatest number of new openings for teachers. Since the southwestern and mountain states are growing considerably in population and school enrollments are on the rise, the need for teachers in those areas is growing at a faster rate than anywhere else in the country.

## Who the Employers Are

**PUBLIC SCHOOLS**
Secondary positions: 925,000
Elementary positions: 1,175,000

**PRIVATE SCHOOLS**
Secondary positions: 93,000
Elementary positions: 184,000

California leads the country in the number of private schools with 2444. Of these, 1574 are under the auspices of a religious organization. New York, with 1923, comes in second; 1504 of these schools are church-affiliated.

## How to Break into the Field

Teacher certification requirements vary according to state and are always handled by the state's department of education. The best way to find out how to be certified in a given state is to seek advice from the education department of your college. Many schools of education are members of the National Council of Accreditation of Teacher Education, and graduates of these schools are most likely to be certifiable in every state. Certain education requirements are necessary for teaching in public schools; for teaching in public or most private high schools, you are also required to have majored in the subject in which you wish to teach. Private elementary and high schools usually do not require either certification or education courses.

College placement offices provide the most information and the best service in the quest for a teaching position. Local school systems keep them informed of openings, and more distant job opportunities can be found on their bulletin boards as well. Information can also be found in the classified sections of local newspapers.

An interesting cover letter and résumé addressed to a principal will usually produce a response or an application form. Visiting the community in which you want to find a job can produce contacts and word of mouth recommendations.

Personal recommendations are always a good way to secure a job, particularly in an independent school. Application to an independent school is made to the headmaster through a letter of introduction, and a résumé that should include any extracurricular activities that would make you a more valuable staff member. The best time to apply is in the fall before the year in which you wish to be employed. Hiring decisions are usually made in March or April.

To save time and gather information about out-of-town schools, it can be useful to contact national placement organizations.

Among these are:

North American Educational Consultants
P.O. Box 995
Barre, VT 05641
802-479-0157
(specializes in secondary schools here and abroad)

Careers in Education
P.O. Box 455
East Stroudsburg, PA 18301

Independent Education Services
80 Nassau Street
Princeton, NJ 08540
800-257-5102 (toll free) or 609-921-6195
(specializes in placing prospective private school
     teachers)

Independent Educational Counselors Association
P.O. Box 125
Forest, MA 02644
(specializes in independent schools for the learning
     disabled)

## International Job Opportunities

Opportunities to teach abroad in a variety of subject areas are available to elementary and secondary school teachers through a program sponsored by the United States Information Agency. An applicant must have at least a bachelor's degree, be a United States citizen, and have three years of successful full-time teaching experience, preferably in the subject and at the level for which an application is made. Information and application materials may be obtained from:

Teacher Exchange Branch
United States Information Agency
301 Fourth Street, S.W.
Washington, DC 20547

Opportunities to teach abroad aren't limited to those with three years of teaching experience. For example, speaking Russian can lead you to be sponsored by the American Field Service to participate in an exchange program with the Soviet Union. The Peace Corps, the National Education Association, and UNESCO all offer opportunities for teachers abroad. The most comprehensive brochure on the subject, *Study and Teaching Opportunities Abroad* by Pat Kern McIntyre, can be obtained from the U.S. Department of Education, Washington, DC 20402.

# TEACHING

## Qualifications

*Personal:*  A genuine desire to work with and care for young people. Ability to lead a group. Strong character. Stamina. Creativity. Well organized.

*Professional:*  College degree. For public schools, certification as well. High school teachers are usually required to have majored in the subject they choose to teach. For elementary school teaching you need a broad range of knowledge and interests, including some instinct for children and how they develop.

## Career Paths

| LEVEL | JOB TITLE | EXPERIENCE NEEDED |
|-------|-----------|-------------------|
| Entry | Teacher | College degree; certification as required |
| 1 | Teacher (with master's) | Master's degree in education or in subject area |
| 3 | Department Head (high school) | 7-10 years |

## Job Responsibilities

## Entry Level

THE BASICS:   Presenting subject matter. Developing lesson plans. Preparing and giving examinations. Arranging class and individual projects that contribute to the learning process. Attending parent conferences, field trips, and faculty meetings. In junior and senior high schools, homeroom guidance, study hall supervision.

MORE CHALLENGING DUTIES:   Club leadership. Sports coach or leading support for sports activities. Directing activities in which the entire school participates, such as assemblies or fund raisers.

## Moving Up

Opportunities for advancement exist in the field of education, especially for those with energy, ideas, and the ability to communicate with both adults and children. Administrators, principals, and superintendents can earn up to $45,000 per year and have a major influence on the communities they serve. To move into administrative or supervisory positions, you must have one year of graduate education, several years of classroom experience, and sometimes a special certificate, depending on the state.

The concept of master teacher is a new one, but school systems in Texas are already awarding teachers who achieve this level—by virtue of experience and effectiveness—with greater responsibility, status and a significant increase in pay. High school teachers have the opportunity to become department heads.

Good and creative teachers with some years of experience are in an excellent position to become educational consultants either as editors of textbooks, directors of special programs, or within the school system as curriculum developers. Especially in the areas of math and the sciences, experienced teachers are being pulled from the classroom to teach other teachers.

To move up in the field of education, you have to have the ability to deal with all kinds of people in all kinds of sensitive situations.

You must have a dedication to the job that continues long after school's over. And you must show a willingness to continue your own education, gaining a master's degree or even a doctoral degree in education or in your subject area.

Teaching can also be a springboard into other professions. Businesses have long known that the experience one gets as a teacher is excellent training for executive positions in marketing, public relations, and advertising. The analytical skills that teachers develop in the classroom, as well as the ability to deal with many different people at many different levels, cannot be taught in a business school. These skills that teachers use day in and day out can be effectively transferred to the business world.

# ADDITIONAL INFORMATION

## Salaries

Teachers progress in salary as they gain experience. Salaries vary widely from state to state. Starting annual salaries can range from $10,000 to $16,000, but generally higher salaries are to be found in the suburban areas. According to the National Educational Association, for secondary and elementary teachers, in 1982-83, the state with the highest average annual salary was Alaska at $33,953. Washington, DC, came in second with an average salary of $26,045. The lowest paying states were Mississippi, with $14,285; Arkansas, with $15,176; and Vermont, with $15,338.

The salary of an independent secondary school teacher is not commensurate with that of a public school teacher, although there are fringe benefits that can often compensate for the difference in pay. At boarding schools, teachers can expect to receive free room and board or rent-reduced housing on or near campus. Travel expenses and smaller workloads also help to compensate for smaller salaries.

## Working Conditions

*Hours:*    Most teachers spend between six and seven hours a day with 15 to 30 children, but that is far from the end of the day. Experienced teachers can often get by with a quick review of their old lesson plans, but new teachers should anticipate an extra two or three hours a day for preparation and correction. Some of the work can be done during the summer vacation and the three or four weeks of holiday during the year.

*Environment:*    This can range from the rolling hills of a rural boarding school to the bleak insides of an inner-city school. Teachers can have access to tennis courts, swimming pools, even a stable full of horses, or they may have to content themselves, for recreation, with the smoke-filled faculty room.

The classroom environment is generally what you make it. Elementary teachers can plaster their walls with children's art, waxed leaves in fall, cutouts of flowers in spring. In secondary schools, decoration can be anything from a chart of chemical elements to a poster of the rock star of the moment.

*Workstyle:*    Most classrooms consist of a blackboard, a big desk, and 20 or 30 smaller ones facing front. Yet each teacher has the choice to make each class either student-centered or teacher-centered. Some teachers prefer to lecture or read to their students; other promote student discussion to various degrees. History teachers have discovered that role-playing and mini-dramas can give life to an epoch. English classes can concentrate on grammar, literature, or writing. With the availability of computers, students are less likely to tune out of math class because of the possibilities for direct interaction and feedback. The art room can offer the freest environment. Many teachers turn on music while students paint or sculpt, and students chat among themselves or with the teacher as they work. The major environmental factor in any class will always be the moods and attitudes of the students, and this is where the good teacher becomes the chief architect of his or her surroundings.

*Travel:* If travel is your end, teaching abroad can be your means. Class trips and outings are also available to the enterprising teacher. Schools will often pay the expenses of a teacher attending a conference, chaperoning a team to an out-of-town match, or accompanying students to recreational activities.

## Extracurricular Activities/Work Experience

Volunteer—Big Brother/Sister program, tutoring, sports, summer camps, teen counseling, child care centers for retarded or culturally disadvantaged children

Athletics—sports (participation can lead to coaching positions in secondary schools)

School—cheerleading, debate society, literary clubs, student newspapers, yearbook publications, student government, drama club, glee club, art club, alumni/admissions administrative work

## Internships

Aside from the student teaching that accompanies college or graduate certification and degree programs, there are few opportunities for teaching internships, except at a limited number of independent schools. They offer the opportunity to get experience in teaching without taking on the responsibility of a full-time teacher. Under the tutelage of a head teacher, the intern learns the ins and outs of the profession. Both school and intern benefit from such a program and the intern is often paid a nominal salary. Contact individual independent schools to see if they have a program.

# Recommended Reading

**BOOKS**

*Don't Smile Until Christmas: Accounts of the First Year of Teaching*, Kevin Ryan, ed. University of Chicago Press: 1970

*The Teacher Rebellion* by David Selden, Howard University Press: 1970

*Teaching School: Points Picked Up* by Eric Johnson, Walker & Company: 1979

**PERIODICALS**

*Academic Journal: The Educator's Employment Magazine* (biweekly), Box 392, Newton, CT 06470

*American Education* (ten times a year), U.S. Department of Education, 400 Maryland Avenue, S.W., Washington, DC 20202

*Arithmetic Teacher/Mathematics Teacher* (ten times a year), 11906 Association Drive, Reston, VA 22091

*The Association for School, College and University (ASCUS) Staffing Annual: A Job Search Handbook for Educators* (annual), Box 411, Madison, WI 53711

*Chronicle of Higher Education* (weekly), 1333 New Hampshire Avenue, N.W., Washington, DC 20036

*Harvard Educational Review* (quarterly), Graduate School of Education, Harvard University, 13 Appian Way, Cambridge, MA 02138

*Today's Education* (quarterly), National Education Association, 1201 Sixteenth Street, N.W., Washington, DC 20036

## Professional Associations

American Federation of Teachers
11 Dupont Circle, N.W.
Washington, DC 20036

Association for School, College and University Staffing
ASCUS Office
Box 411
Madison, WI 53711

National Education Association
1201 Sixteenth Street, N.W.
Washington, DC 20036

# INTERVIEWS

**Margaret Thompson, Age 39**
**Secondary School Teacher**
**Needham High School and Newman Middle School, Needham,**
**    MA**

Although the competition for English teaching positions is very fierce now, in five or ten years there should be a real market. If you're lucky enough to find a position right now, though, it can be the most rewarding experience of your life.

There's autonomy. You're your own boss. Once you're in that classroom, the door is shut. You can be amazingly creative. Every single day is a new challenge. And you get to show off a lot. Probably more than any other subject, teaching English calls on you to perform.

Teaching English is like conducting an orchestra. You're in charge, but the students have to play their own instruments. The more they play their instruments, the happier they are. And when they can hear those instruments being played together in an orchestra it gives them a big thrill.

For example, sixth and seventh graders are being introduced to literature for the first time in their lives; you teach them about the components of a story—the climax, the development, the resolution; then they read the story and for the first time have a sense of recognition; they're reading about characters who are experiencing some of the same conflicts they are. They listen to each other's interpretations and all of a sudden they don't feel alone anymore.

I think literature is one of the most subtle ways of affecting people's lives. You have to know how to choose the literature that will be the catalyst to self-understanding and give students the structural tools to explore their own lives.

What kind of person would really have fun teaching? A caring person. One who enjoys stroking and being stroked. I've been teaching for 13 years now. I'm making about maximum salary, $26,000 a year. But the feeling of loving more than balances the salary. I'll never stop teaching. It's a joy.

**Laura Daigen, Age 29**
**Fourth-Grade Teacher**
**New York, NY**

I started out teaching Spanish in a high school, but I decided there was a lot more good to be done as a bilingual elementary teacher. I was somewhat disillusioned with the politics of the school where I worked. Teachers who had been there longer made sure that they had no problem students in their classes and that kids with low test scores were taught by the rookies. The Spanish language textbooks were poor and antiquated in their absence of girls and women.

When you're working in a public school in the inner city, the needs of your kids are infinite. There's very little support from the administration or cohesiveness among the faculty. You're alone with kids who have more emotional needs than you can deal with. There is no one to help you help them.

Even under the best of conditions, the hours are long. Sometimes I'm there until six or seven and I'm not the last one to leave. I'm making plans for new curriculum, I'm marking papers, I'm cleaning up the room. You're never finished when you're a teacher: you dream it, you go to sleep thinking about it, and you wake up thinking about it. You're teaching eleven pieces of curriculum and you have to coordinate them and give enough time to each one. It was easier teaching Spanish in a high school, because I was teaching one subject at three different levels and the students had the responsibility of absorbing what you gave them. But when you're teaching elementary school and the kid isn't learning, it's your fault.

If someone were trying to decide whether or not to go into teaching, the first question I would ask him or her is: Do you like kids? Are you ready to take on the responsibility of little human beings and their emotional needs? There is tremendous satisfaction in watching children grow, seeing them move out of themselves and being able to give; recognizing their right to receive fair treatment knowing they can effect change if they see an injustice being done. It feels great to have a kid say, "I like math this year," or "I don't have to try to read these books, I can just read them," or to see kids take charge of their own lives and take pride in what they've done on their own.

I think what makes a good teacher is the sincerity of his or her commitment—whatever it is. It can be to help children love reading or to help them see themselves as important pieces of society or to help them realize the value of their own ethnicity. It has to be more than a vague "I love kids"—but you have to love kids, too.

No matter what I end up doing with my life, I'll always be able to look back on these five or six years and say those were years that I spent doing what I really wanted to do. Perhaps I haven't been well-compensated economically, but the job is rewarding. I will always be able to say that I did something that was meaningful, something that helped others.

# INSURANCE

EVER since James Bolter, a resident of Hartford, Connecticut, paid a two-cent premium for a $1000 policy to the Travelers Insurance Company in 1864, protecting him against physical mishaps on his daily walk to the post office a few blocks away, Americans have been insuring themselves and their possessions. In fact, about 90 percent of Americans today have some form of insurance—health, life, home, or auto.

The scope and direction of American insurance companies have changed considerably since Mr. Bolter paid his two pennies. Today the industry, usually divided into two sectors—health and life, and property and casualty—is concerned not only with individuals but with large and small groups of people as well. It offers protection for almost any condition that might arise, including industrial pollution, nuclear accidents, malpractice, kidnapping, crop failures, and libel.

Despite trying economic conditions during the early 1980s, the need for insurance coverage continued to grow. In the property and casualty sector, approximately $104 billion was spent for premiums, and this figure is expected to reach more than $300 billion by the year 2000.

At present, Americans own more than $3 trillion worth of life insurance. They and their employers pay more than $111 billion per year for health care coverage. What this all adds up to is a very healthy people-oriented industry with many career opportunities for talented, enterprising, and industrious individuals.

Even during the recessionary period of the early 1980s, employment in the industry increased approximately 4 percent, without taking into account an estimated 270,000 people in 1982 who worked as independent agents or were staff employees of such agencies. Looking at casualty/property and health/life separately, the employee growth rate was 20 percent and 8 percent respectively for the period between 1981 and 1982.

Although technology as well as the faltering economy has been responsible for a "compression" in number of clerical and secretarial jobs, the need remains strong in the following areas:

- **Sales**
- **Actuarial**
- **Underwriting**

This is partly because Americans tend not to cut back on insurance coverage even during recessionary periods, and because there will be more Americans between the ages of 25 and 55 who will need coverage.

There is encouraging news for women seeking careers in the insurance industry, too. Data from the Equal Opportunity Commission show that 42 percent of all professional positions in the insurance industry in 1981 were held by women. In 1970, it was 16.9 percent. A total of 26 percent of employees classified as managers or company officials in 1981 were women; only 11 percent of those positions were filled by women in 1970.

## Job Outlook

*Job Openings Will Grow:*   As fast as average

*Competition for Jobs:*   Average

Because insurance is not a glamour industry, there is not usually an overabundance of applicants. Many major insurance companies are eager to train ambitious liberal arts or business administration majors at company expense.

*New Job Opportunities:* Because government deregulation has allowed many insurance companies to offer financial services as well as insurance, individuals with a background in financial planning will be needed to advise and counsel clients. Also, as insurance becomes perhaps the most important of all fringe benefits employees receive, more companies will use the services of insurance consultants and financial planners to select insurance, annuities, and pension plans.

## Geographic Job Index

Although insurance agencies are found in almost all communities, large or small, major insurance companies are primarily located in the eastern corridor—Boston, MA, New York, NY, Philadelphia, PA, and Washington, DC. Other U.S. cities with a large concentration of insurance companies are Milwaukee, WI, Chicago, IL, Indianapolis, IN, Dallas, TX, and San Francisco, CA.

## Who the Employers Are

The best chances for entry-level positions are often at the home offices of large firms where extensive training programs are available for newcomers. Many of these firms actively recruit on college campuses and advertise in the annual *Insurance Careers,* a magazine supplied to college placement offices.

*Major Employers*

Aetna Life and Casualty Company, Hartford, CT
Allstate Insurance Company, Northbrook, IL
Equitable Life Assurance Society, New York, NY
Group Health Association of America, Inc., Washington, DC

Hartford Insurance Group, Hartford, CT

Insurance Company of North America, Philadelphia, PA

Liberty Mutual, Boston, MA

Massachusetts Mutual Life Insurance Company, New
York, NY

Metropolitan Life Insurance Company of New York, New
York, NY

Mutual Life Insurance Company of New York (MONY),
New York, NY

Mutual of Omaha, Omaha, NE

New England Life, Boston, MA

The Prudential Insurance Company of America, Newark,
NJ

State Farm Insurance Company, Bloomington, IL

Travelers Insurance Company, Hartford, CT

## How to Break into the Field

If you are still in school, check with your school's placement office about insurance recruiters who regularly visit college campuses. Many graduates who are offered employment in the industry make their initial contact with the company during a campus visit by a recruiter. Another suggestion is to check the classified advertisements of your local insurers. Also, don't neglect contacts you might make through summer or part-time work with an insurance company. Several large companies offer internships or have cooperative education agreements with colleges. The best way to find out about them is to write directly to the companies that interest you.

There are also several courses in insurance that you might consider taking. In addition to any offered on your campus, you might want to consider specialized or part-time programs offered by:

The College of Insurance
123 William Street
New York, NY 10038

(This accredited college, supported by the insurance industry and the Insurance Society of New York, offers a nine-month introductory course. It also offers specialized courses in its night division.)

The American College
270 Bryn Mawr Avenue
Bryn Mawr, PA 19010

(This college is responsible for the more than 250 programs across the United States that offer certified life underwriter (C.L.U.) and chartered financial planner (Ch.F.P.) certified programs.)

American Institute for Property and Liability Underwriters
Insurance Institute of America
Providence and Sugartown Roads
Malvern, PA 19355

(This organization administers programs across the United States that offer the chartered property casualty underwriter (C.P.C.U.) designation.)

Professional Insurance Agents
400 North Washington Street
Alexandria, VA 22314

(This professional organization has been offering three-week summer introductory courses in insurance for more than 20 years.)

## International Job Opportunities

If you're working for a large insurance company that (1) is doing business overseas; (2) is looking to do business overseas; or (3) has acquired foreign insurance companies, you might find yourself in a position to accept employment overseas. Usually, individuals who speak a foreign language and work as underwriters or sales agents have the best opportunities to be transferred to their firm's foreign offices. Because the American market, especially in health/life insurance, is saturated, insurance companies are looking abroad to increase their sales.

# **SALES**

Responsible for helping to plan their clients' financial security, sales agents are the backbone of the industry. In addition to bringing new business to the company, they often help clients file claims and keep them informed about new or better insurance options. They are also responsible for encouraging the client to renew his or her policy.

The two types of agents are those who work on contract for a small salary, benefits, and commissions for one company, and those who work exclusively on commission for many different insurance companies. The latter are often referred to as brokers or independent agents.

According to the Bureau of Labor Statistics, approximately one of every four agents is self-employed. Some agents specialize in either health/life or casualty/property. Others sell both types of insurance. It is not uncommon for the independent agent to sell real estate or to advise on special financial options—such as annuities or mutual funds—in addition to selling insurance policies.

Agents must be licensed in the state where they sell insurance. Licensing usually requires that the individuals pass a state examination that tests knowledge of insurance principles and state insurance laws.

## Qualifications

*Personal:*   Drive. Willingness to work hard to develop a clientele. Good appearance. Self-assurance. Friendly, outgoing, not easily discouraged.

*Professional:*   Excellent oral communications skills. Ability to make sales presentations. Understanding of the personal and financial needs of diverse groups of people.

## Career Paths

| LEVEL | JOB TITLE | EXPERIENCE NEEDED |
|-------|-----------|-------------------|
| Entry | Sales trainee, sales representative | High school diploma; college degree helpful |
| 2 | Sales agent | 6-18 months |
| 3 | Sales manager | 3-5 years |
| 4 | District manager | 5-7 years |
| 5 | Regional manager | 7-10 years |
| 6 | Vice president, sales | 10+ years |

## Job Responsibilities

### Entry Level

THE BASICS:Learning about the business of selling insurance. Attending sales strategy sessions as an observer or "tailing" an experienced agent on his or her calls. Assisting established agents to service accounts.

MORE CHALLENGING DUTIES:Making sales calls. Servicing new accounts. Finding and soliciting potential clients.

### Moving Up

Successful salespeople often rise quickly in the insurance business. Sales careers can lead to positions as regional, district, and general managers, as well as corporate vice presidencies. In order to move up you must not only show creativity and ability in selling and servicing your accounts, you must also demonstrate your potential as a manager. To move into the corporate structure you must acquire a broad understanding of all functions of the company and of the insurance industry in general.

# ACTUARIAL

If salespeople are the backbone of the insurance industry, then actuaries are the brains that make the organization function successfully. These individuals determine premiums and contract provisions of policies. They also calculate the probability of loss due to death, natural phenomenon, sickness, fire, theft, or other catastrophes.

According to the Bureau of Labor Statistics, this profession will see a faster than average growth rate during the 1980s and 1990s. Consequently, many opportunities are available for the individual who strives for challenges and success.

Being an actuary is not a career for anyone with a fear of math. In fact, insurance companies expect actuarial trainees to have been math, business administration, or economics majors and to have taken courses in probability, statistics, and calculus. Computer science course work is a definite plus. Some insurance companies also hire engineering or science majors who have a strong background in mathematics (usually between 20 and 25 credit hours).

Many insurance companies expose their actuary trainees to several of the firm's operations, providing them with a variety of professional experiences. Actuaries may be found in pension planning, group underwriting, investment, or other departments.

An actuary trainee is expected to study for a series of examinations, either individually or through specific courses given by the insurance company or the continuing education division of a nearby college or university. It usually takes between five and ten years to complete the ten examinations, and it is not uncommon for the trainee to study more than 25 hours a week for the examinations. When you have passed all the necessary examinations you become a fellow of the sponsoring organization—either the Society of Actuaries or the Casualty Actuarial Society. This designation marks you as a full-fledged member of the profession, much the same way that passing the bar attests to your professional status as a lawyer. Unlike law, which you cannot practice without passing the bar, you do not need the fellow designation to do the work of an actuary, but you must have it in order to progress in the field. An increasing number of employers hire actuary trainees who have already successfully completed first- and second-level examinations while still in college.

## Qualifications

*Personal:* Patience. Detail-oriented. Ability to work well individually and on a team.

*Professional:* Strong math and business orientations. Ability to present information in an orderly fashion. Computer skills.

## Career Paths

| LEVEL | JOB TITLE | EXPERIENCE NEEDED |
|-------|-----------|-------------------|
| Entry | Actuary trainee | College degree with strong math background |
| 2 | Actuary | 1-3 years; at least two professional examinations passed |
| 3 | Actuary, | 3-5 years; at least five junior associate exams (Society of Actuaries) or seven exams (Casualty Actuarial Society) passed |

| 4 | Actuary, associate | 5-8 years; six or more exams passed |
| 5 | Actuary, new fellow | 8-10 years; all ten exams passed |
| 6 | Actuary, fellow | 10+ years |

## Job Responsibilities

### Entry Level

**THE BASICS:**   Learning how the firm operates through a rotation of training assignments. Working on the simpler actuarial tables. Solving problems contained in them.

**MORE CHALLENGING DUTIES:**   Applying mathematical models to a variety of anticipated health or casualty losses. Designing insurance proposals. Calculating premiums for a new type of insurance. Determining what benefits a policy should contain. Providing statistical information at government hearings on insurance rate changes. Preparing presentations for clients.

### Moving Up

It is impossible to progress in this field without passing the examinations given by the Casualty Actuarial Society (for casualty/property insurance) or the Society of Actuaries (health/life insurance). The sooner you attain the title fellow, the sooner you will be ready to enter the managerial ranks. You must also prove your abilities by working on more complex tables and presentations and by being able to arrive at workable solutions to more and more complicated problems. The career opportunities for an actuary are considerable—the presidents of several major insurance companies started their careers as actuaries.

# UNDERWRITING

Whereas the sales force acts as the backbone and the actuaries the brains, underwriters are the legs and arms of the organization. They are responsible for determining insurance risk—should Jane Smith or the members of Bill Doe's group be insured, and if so, for how much? What additional risks do these people have?

Although underwriters often have a low profile in many insurance companies, their importance cannot be overemphasized. For example, if policies are priced too low, the insurance company could stand to lose money. Similarly, if policies are priced too high, the insurance company could lose business to its competitors.

Most underwriters work in the home offices of their companies. In some organizations, they provide information regarding specific policies to independent or in-house brokers. Advancement often depends on the successful completion of a series of C.P.U.C. (certified property casualty underwriting) examinations. At the entry level a business administration degree is helpful, although many liberal arts majors are hired.

## Qualifications

*Personal:* Ability to work well with others. Patience. Attention to detail and follow-through.

*Professional:* Aptitude for figures and an understanding of business dynamics.

## Career Paths

| LEVEL | JOB TITLE | EXPERIENCE NEEDED |
|-------|-----------|-------------------|
| Entry | Trainee | College degree |
| 2 | Junior Underwriter | 1-5 years and continuing education |

| 3 | Underwriter | 5+ years |
| 4 | Underwriting supervisor manager | 7+ years |
| 5 | Vice president | 10+ years |

## Job Responsibilities

### Entry Level

**THE BASICS:**   Evaluating routine applications under the supervision of an experienced underwriter.

**MORE CHALLENGING DUTIES:**   Evaluating applications without supervision. Working on more complex applications.

### Moving Up

It will be necessary to earn the fellow designation from the Life Underwriting Training Council in order to move up in this field. Because you will be making decisions about who is to be insured and at what rates, your knowledge must keep up with changes in the industry so that you can make good decisions. With increasing knowledge and experience, you will also be asked to handle larger, more complex groups. Underwriting supervisors frequently write company policy, so a thorough knowledge of company philosophy and procedure will be required before you can move into a managerial or corporate position.

# ADDITIONAL INFORMATION

## Salaries

**SALES**

Initially, a small salary ($10,000 to $12,000 a year) plus commissions or other incentive is offered to the tyro salesperson. Later (the length of time varies depending on the company or agency but can be anywhere from 6 to 30 months), it can be sink or swim with commissions only. Some insurance companies, however, still continue to provide small salaries for their agents.

Annual compensation for successful agents can be excellent. For example, life insurance agents with five or more years of experience had a median income of $35,000 a year in 1982. Many agents, especially those who sold casualty policies, earned upwards of $50,000 annually.

## ACTUARIAL

Starting salaries for entry-level positions vary, often depending on geographic locations as well as on whether the trainee has already passed any of the three lower-level examinations jointly sponsored by the professional actuary societies. Typical 1982 starting salaries for an actuary trainee, who had not completed any exams, was between $16,000 and $17,000 annually. After successfully passing one examination, an actuary earned between $17,000 and $18,500 a year, after passing two examinations, between $18,500 and $20,000 a year.

Depending on the number of professional examinations completed, salaries rise appreciably. For example, average annual salaries of new associates of both actuary societies was between $24,000 and $28,000 in 1982. Actuaries who became fellows in that same year had yearly salaries between $35,000 and $45,000. Fellows with additional experience are known to earn upwards of $50,000 annually.

## UNDERWRITING

A salary survey conducted by several property and liability insurance companies found that the median salary for underwriters in 1982 was $18,500 a year. Senior underwriters working with personal lines of insurance had median annual salaries of $21,000 in comparison to commercial line underwriters who earned $23,700. Supervisors in property and liability companies had incomes between $26,500 and $28,000 a year.

## Working Conditions

*Hours:*   Expect work weeks of between 35 and 40 hours, depending on the firm. Overtime is usually not mandatory for underwriters and actuaries except during certain busy periods.

For sales agents who have passed the training period, things are a bit different. Often they make their own hours, scheduling appointments around a client's time schedule. This can mean night and weekend hours as well as more than the usual 35-to-40-hour work week.

*Environment:*   The quality of work space varies greatly, depending on the number of people in the office. Don't expect your own office as a newcomer. More than likely you'll have to share space bullpen-style with co-workers. Private offices usually go to more senior staff members.

*Workstyle:*   Better get used to working at your desk. With the exception of sales agents, who must interact with clients in their homes or places of business, individuals working in the insurance industry perform many of their tasks on computer terminals, typewriters, and other office equipment located adjacent to their desks.

*Travel:*   In general there is little travel at the entry level, except locally. Agents usually see their clients outside the office. Underwriters and actuaries sometimes travel to branch offices. Associate and senior-level employees attend industrywide conferences and seminars.

### Extracurricular Activities/Work Experience

Serving as club treasurer

Working on fund-raising drives

Student membership in professional organizations

Part-time or summer jobs in sales, in the
benefits/personnel departments of corporations
or in insurance companies.

## Internships

Many companies offer internships and cooperative work-study
arrangements. No single directory lists all these opportunities, but
your school placement service can be of assistance. You can also
write directly to the personnel departments of companies that
interest you.

## Recommended Reading

**BOOKS**

*Glossary of Insurance Terms* by Thomas E. Green, Robert
W. Osler, and John S. Bickley, The Merrit Company: 1980

*The Insurance Almanac*, Underwriting Publishing Company:
1982

*Insurance Facts*, Insurance Information Institute: 1983

*Life Insurance Factbook*, American Council of Life Insurance: 1983

*Principles of Insurance* (7th edition) by Robert I. Mehr and
Emerson Cammack, Richard D. Irwin, Inc.: 1980

*Principles of Insurance* by George E. Rejda, Scott, Fores-
man: 1982

**PERIODICALS**

*Best's Review* (monthly), A.M. Best Company, Inc., Ambest
Road, Oldwick, NJ 08858

*Business Insurance* (biweekly), Crain Communications, 740
North Rust Street, Chicago IL 60611

*Insurance Careers* (annual), Resort Managements, Inc., P.O. Box 4169, Memphis, TN 38104

*Life Insurance Selling* (monthly), Commerce Publishing Company, 408 Olive Street, St. Louis, MO 63102

*National Underwriter* (weekly), National Underwriter Company, One Marine View Plaza, Hoboken, NJ 07030

*Professional Agent* (monthly), Professional Insurance Agents, 400 North Washington Street, Alexandria, VA 22314

*Today's Insurance Woman* (quarterly), National Association of Insurance Women, P.O. Box 4410, Tulsa, OK 74159

(For a comprehensive guide to insurance periodicals, check the *Insurance Periodicals Index,* issued annually by the Special Libraries Association and NLS Publishing Company.)

## Professional Associations

American Council of Life Insurance
1850 K Street, N.W.
Washington, DC 20006

American Institute for Property and Liability Underwriters
Insurance Institute of America
Providence and Sugartown Roads
Malvern, PA 19355

American Insurance Association
85 John Street
New York, NY 10038

American Society for Chartered Life Underwriters
270 Bryn Mawr Avenue
Bryn Mawr, PA 19010

Casualty Actuarial Society
One Penn Plaza
New York, NY 10001

Independent Insurance Agent of America, Inc.
85 John Street
New York, NY 10038

Insurance Information Institute
110 William Street
New York, NY 10038

Insurance Library, The College of Insurance
123 William Street
New York, NY 10038

Insurance Library Association of Boston
156 State Street
Boston, MA 02109

Life Underwriting Training Council
1922 F Street, N.W.
Washington, DC 20006

National Association of Casualty and Surety Agents
5225 Wisconsin Avenue, N.W.
Washington, DC 20015

National Association of Independent Insurances
2600 River Road
Des Plaines, IL 60018

National Association of Insurance Women
P.O. Box 4410
Tulsa, OK 74159

National Association of Life Underwiters
1922 F Street, N.W.
Washington, DC 20006

Professional Insurance Agents
400 North Washington Street
Alexandria, BA 22314

Risk and Insurance Management Society, Inc.
205 East 42nd Street
New York, NY 10017

Society of Actuaries
500 Park Boulevard
Ataska, IL 60143

Society of Chartered Property and Casualty Underwriters
Kahler and Hall
Providence Road
Malvern, PA 19355

# INTERVIEWS

**Maria Leon, Age 25**
**Analyst**
**Royal Insurance Company, New York, NY**

Unlike many analysts and underwriters, I did not have a liberal arts background. In fact, I knew when I was planning for college that I wanted a career in insurance and wanted to remain in New York City. Therefore, I chose the College of Insurance and participated in their five-year co-op program and earned a B.B.A. degree in business administration.

This co-op program enabled me to work on a part-time basis with Royal. I would go to school for a semester and work the following semester. Royal paid two-thirds of my tuition during college semesters and a regular salary when I worked there in alternate semesters. The experience was great—I was able to work in the international, marketing, claims, and underwriting departments.

As a full-time analyst with Royal for the past one and a half years, I work closely with underwriters. I have earned my C.P.C.U. (casualty property certified underwriter) designation, the same as any underwriter would. This is similar to an advanced degree and increases knowledge as well as the value of the underwriter to the organization.

And just what does an analyst do? Basically, I service the underwriters. This means I present them with research data-specific industry reports, or even new information about different types of insurance, and help them compare and select risks. Normally we'll write a company manual that details what the underwriters should look for when determining rates and risks.

In the research capacity, one of my responsibilities is to keep track of legislation in my insurance area—commercial lines. What changes in California vehicle laws might make for a better risk, or under what circumstances a policy may be cancelled—this is the type of information I collect.

During a typical day I'll work on short- as well as long-term research projects, and I'll often take calls from branch offices regarding procedures and forms in addition to answering some underwriting questions. Sometimes someone in the branch office will ask for an endorsement (a change in policy) to cover a special situation. What I'll do is try to phrase it correctly and then pass it along to our legal department. If it's acceptable, I'll notify the individual in the branch.

It is not uncommon for an analyst to travel to branch offices for research. Last year I spent some time in our Atlanta, GA, office, looking through their automobile underwriting files. This research gave me a better perspective on what actually was happening in the marketplace, and this is the type of information that would influence future policy changes or rate hikes.

**Gil Goetz, Age 26**
**Underwriter**
**Kemper Group, New York, NY**

After I graduated from Ohio Wesleyan University in 1980 with a degree in economics, I decided to take advantage of an opportunity to learn insurance brokering in London, England. There, I became a broker trainee. I returned to the United States within a year.

I had heard about Kemper's underwriting training program and wanted to become an underwriter, so I applied and was accepted as a trainee. My training included spending a month at the Kemper home office in Long Grove, IL, where I received excellent, intensive training.

As a casualty underwriter, I represent the company to brokers and agents. I answer questions about policies and often recommend the policies that might be best for a client's needs. I also service these policies, which means that I provide the brokers with the information they need to attract and maintain business.

I also examine potential insureds. In examining these accounts, it is essential that the underwriter look at inspection reports and research. For example, if a broker calls and wants to know what rate we will charge to insure a large manufacturing plant, I'll look at an inspection report made out by an engineer and at other data available about the type of work done at the plant. If an underwriter underestimates the risk, the insurance company stands to lose money; if the risk is overestimated, the client will take its business elsewhere.

We're using computers more and more in our operations. If a broker calls, we can check to see if premiums are still outstanding for the client and find what endorsements might have been made. You are on the telephone a lot in this job. You have to know where to find information and be able to provide it quickly.

I enjoy working with people, and I like the feeling that I am providing a service to society. I get a lot of satisfaction out of doing my job.

# MARKET RESEARCH

I N ancient times, people flocked to star-gazers and oracles to ask about the future. Today predictions are made by market researchers, who try to foresee if a particular group—consumers, businesspeople, voters—will believe in a product, a business decision, a candidate, or an idea. Rather than consulting the stars or relying on their instincts, market researchers base their predictions on research. They spend days, weeks, months, and even years sampling opinions, tastes, and reactions.

Many variables besides quality account for success or failure in the marketplace—changing tastes and trends, public perceptions (and misconceptions), subliminal messages. Before a new product appears in the stores, before an old favorite gets new packaging, even before a product is discontinued, market research is conducted and its conclusions incorporated into business strategies and advertising campaigns.

Food processors and manufacturers of cosmetics and consumer goods need to know who will buy a certain product and why, what kind of packaging is most enticing and appropriate, how much consumers are willing to pay, and how frequently consumers will use the product. To find the answers, new products are tested against established ones, facsimiles of the product are test-marketed, and surveys of potential consumers are made. The results may confirm a company's conviction that a sufficient market for a given product exists or that it is the right time to introduce such a product. Market researchers often suggest pricing and marketing strategies based on consumer responses.

Research may be concerned with something other than a consumer product. Current issues, the ramifications of business decisions, and even political campaigns are studied through market research. Regardless of the subject, all research follows a similar pattern—the problem is defined, a research strategy is developed, information is collected and analyzed, and the results are interpreted and presented to the users. Information may be gleaned from phone interviews, on-the-scene impromptu interviews with a target group (e.g., moviegoers, shoppers), written surveys or focus groups, the carefully selected people who fit the demographics of the client's intended market.

Two different types of work go into market research: data collection and information analysis. Data collection is the more detailed. It involves tallying questionnaires, conducting phone interviews, and assembling relevant printed information. Much of the research is either quantitative or qualitative. Quantitative research tends to produce numbers-oriented studies; qualitative reports feature more subjective information, such as opinion sampling. Information analysis is more sophisticated work; it involves interpreting the results of the research and writing up conclusions in a report or presentation for the client. There is also tedious number-crunching work, which is given to the least experienced people. But computers are eliminating the drudgery of such work and, best of all, drastically reducing the amount of time needed to do complicated cross-tabulations.

Some companies do both information collection and analysis, but market research firms often rely on tabulation services and field services to perform the time-consuming tasks that precede analysis. Tabulation services specialize in collecting data of any sort, usually through surveys and questionnaires supplied by the research company. They do not analyze the data, they simply collect it. Field services perform a more personalized aspect of the same task. They specialize in interviews that take many forms: with groups, with individuals, by phone, or in shopping centers. These companies may tabulate the results for their clients.

Market research is a growing field that is extending its influence into businesses of all kinds and to hospitals, colleges, and other nonprofit organizations that need to know how to manage their own development, who their constituency is or ought to be, and, often most important, how to target potential contributions and successfully solicit donations.

No one specific academic background is required for a job in market research, but you must be able to work effectively and comfortably with numbers and to articulate your ideas clearly and convincingly. Although an undergraduate degree is all you'll need to get an entry-level position, you should be prepared to take some graduate courses if you want to make your career in this field. Although you don't have to be a math major to do quantitative research, you will probably feel more comfortable in that area if you have a strong background in math or statistics. After you gain some experience and demonstrate your ability to grasp quickly what kind of research is necessary in any particular client situation, you should be able to move up into a challenging position.

## Job Outlook

*Job Openings Will Grow:*  Faster than average

*Competition for Jobs:*  Keen

*New Job Opportunities:*  Expansion in the industry is taking place in research companies—independent firms that offer research services to clients. As new service industries, such as the newly expanding banking industry and cable television, continue to enter the marketplace, the need for research companies will grow.

## Geographic Job Index

New York, NY, and Chicago, IL, are the leading centers of market research, but job opportunities exist across the country, mostly— but not exclusively—in urban areas.

## Who the Employers Are

MARKET RESEARCH FIRMS perform complete research studies, although they may use outside tabulation houses and field services for data collection. Once they receive the data, they analyze and report on the data to their clients. They may also initiate their own research projects and then sell the findings to interested companies. Research companies often specialize in a single type of client. As a result, many firms have staffs of fewer than ten people.

CONSUMER GOODS MANUFACTURING FIRMS often have market research departments within their corporate structure. These departments may or may not carry out numerical research, but they design the research projects that need to be done, analyze and interpret the results, and make recommendations based on their conclusions. Usually, only a handful of people are employed in-house, and most of the jobs there go to experienced people.

*Major Employers*

> A.C. Nielsen Company, Northbend, IL
> Arbitron Ratings Company, New York, NY
> Audits & Surveys, New York, NY
> Burke Marketing Services, Cincinnati, OH
> IMS International, New York, NY
> Market Facts, Chicago, IL
> Marketing and Research Counselors, Dallas, TX
> NFO Research, Toledo, OH
> NPD Group, Port Washington, NY
> Sellings Areas-Marketing, Inc., New York, NY

## How to Break into the Field

You'll stand a better chance of getting hired in an entry-level job if you have related experience. Field services employ large numbers of high school and undergraduate students in part-time and summer positions, usually as interviewers who call people across the country to ask prepared questions and take responses.

Large firms (those with 40 employees or more) occasionally recruit on campus, but, in general, you will have to investigate openings on your own. Firms are listed in the Yellow Pages under "Marketing Services." Newspaper help-wanted ads normally advertise positions for experienced personnel only. But it's often worthwhile to contact the company that is advertising the job on the chance that the new senior-level person may need an assistant.

The Chemical Marketing Research Association runs a yearly three- or four-day course in July. It's titled Basic Chemical Marketing Research Short Course, but it is a good introduction to the market research process in general. For information, contact the Chemical Marketing Research Association. (Address given at the end of this chapter.)

## International Job Opportunities

Market research is done in most Western countries, but international job opportunities are very limited, since most positions are filled by nationals.

# RESEARCH ANALYSIS

The entry-level position in research analysis is the junior or associate analyst. You will assist experienced analysts by handling the routine detail work that accompanies all projects. In firms or departments that do field work, you may begin by interviewing or editing and coding—checking to see that all questionnaires are completed and assigning numerical codes to nonnumerical responses so that results can be tabulated. Firms that depend on field services may send new employees to a service for a few weeks so they can get a firsthand understanding of the information-gathering process.

Typically, one analyst is responsible for each project. Besides preparing and presenting the conclusions of a study, this individual must coordinate the efforts of all contributors—one or more junior analysts and field and tabulation personnel. Analysts, especially those in senior or executive positions, may solicit new business and maintain contacts with past clients.

To become a successful market research analyst, you must be able to (1) read a qualitative report and determine what questions need to be asked to get concrete, quantitative answers; (2) produce a logical, coherent picture from the results of a numerical survey or the varied answers of an opinion poll; and (3) communicate effectively, guiding your clients to conclusions that are appropriate to the results of your research, convincing them that your recommendations are well-grounded and credible. Deadline pressure is constant; the entire process of research and analysis often takes no longer than six to ten weeks. You may work on one project at a time or be concerned with several in various stages, and you're sure to have assignments changed suddenly.

## Qualifications

*Personal:*   Good judgment. Able to communicate clearly. Personable appearance and manner. Good powers of concentration. Ability to work on a team. Problem-solving mentality.

*Professional:*   Strong writing and communications skills. Math and statistical know-how. Good phone manner. Typing or word processing skills.

## Career Paths

| LEVEL | JOB TITLE | EXPERIENCE NEEDED |
|-------|-----------|-------------------|
| Entry | Coder-editor, junior or associate analyst | College degree |
| 2 | Analyst | 1-2 years |
| 3 | Senior analyst, research manager | 4-7 years |
| 4 | Market research director | 7+ years |

## Job Responsibilities

### Entry Level

**THE BASICS:** Clerical duties: typing, filing, handling the flow of correspondence. Proofreading questionnaires. Writing cover letters, memos, and progress reports. Organizing completed studies and reports.

**MORE CHALLENGING DUTIES:** Writing questionnaires, using successful examples. Basic data analysis. Organizing and rearranging tables, charts, and raw data. Writing introductions and reports for coders and interviewers.

### Moving Up

The key to success in this competitive business is building an understanding of the research process as quickly as possible. You must demonstrate the ability to study data and apply your conclusions to specific problems. You will not deal directly with the users of the research until you have adequate experience (a year or more). At first, you will meet with brand managers, gathering background information and drafting initial proposals. This work will provide an understanding of user needs. As a researcher handling your own projects, you will respond to these needs through your analyses. A good deal of writing and speaking is involved, and you must be both candid and diplomatic in the way you present results.

# ADDITIONAL INFORMATION

## Salaries

Each year, the American Marketing Association conducts a salary survey of its members. The average annual salaries for market research personnel nationwide in the 1983 report were:

| | |
|---|---|
| Analyst | $24,128 |
| Manager or supervisor | $32,945 |
| Director | $41,361 |
| President | $53,483 |

## Working Conditions

*Hours:*   The standard 40-hour week can stretch to 12-hour days and include weekends in times of heavy business. But there are corresponding slack periods. Hours tend to be more regular in the larger research firms and in-house departments.

*Environment:*   Analysts often have their own small offices, and most research firms have at least one larger conference room where analysts will get together to discuss ongoing projects.

*Workstyle:*   The work day is hectic for both entry-level and senior personnel. The junior analyst generally spends each day in the office. Analysts visit clients and potential clients. Research often involves a great deal of phone work.

*Travel:*   Opportunities for national and international travel exist at many research firms. Although field services can provide coverage of virtually any location in the United States, an analyst may occasionally travel to supervise research for an international corporation on market conditions in another part of the globe, or to investigate personally a locale's unique characteristics.

### Extracurricular Activities/Work Experience

Business Research Practicum

American Marketing Association—student member

Canvassing/phone interviewing for charities or political
    campaigns

Campus newspaper—reporting

### Internships

There are no formal internship programs in market research, but you can try to set one up on your own by contacting the personnel director in some of the larger companies.

## Recommended Reading

**PERIODICALS**

*Careers in Industrial Marketing Research* published by The Chemical Marketing Research Association (free)

*Careers in Marketing* by Neil Hobart, American Marketing Association, Monograph Series #4 (free)

*Employment and Career Opportunities in Marketing Research* published by the Marketing Research Association, Inc. (free)

*Marketing Communications* (monthly), United Business Publications, 475 Park Avenue South, New York, NY 10016

*The Marketing News* (biweekly), American Marketing Association, 222 South Riverside Plaza, Chicago, IL 60606

*Marketing Times* (bimonthly), Sales and Marketing Executives International, 330 West 42nd Street, New York, NY 10036

## Professional Associations

American Marketing Association
250 South Wacker Drive
Chicago, IL 60606

The Chemical Marketing Research Association
139 Chestnut Avenue
Staten Island, NY 10305

The Life Insurance Marketing Research Association
170 Sigourney Street
Hartford, CT 06105

# INTERVIEWS

**Tom Keels, Age 29**
**Research Director**
**Louis Harris Associates**
**New York, NY**

I got into market research in 1980. After working in publicity at a publishing house, I decided to switch careers. I knew I liked marketing, so I looked at advertising, marketing consulting, and market research. A friend was working at a small research company called Crossely Surveys; he helped me get a part-time job as an editor-coder so I could see if I liked the work. I found it interesting, so when I was offered a job as a project assistant a few months later, I accepted.

I started in an entry-level job because very few skills transfer directly from publicity to market research. I had good writing skills and was comfortable talking on the phone, but these were not important immediately because I wasn't in a position to deal with clients.

As a project assistant I carried out the mechanical details of a particular job. In other words, I made sure the questionnaires were typed, proofed, and sent out to the field. I supervised some interviewing and oversaw editing, coding, and computer cross-tabulation. I remember very clearly one of the first projects I worked on. The client was a major manufacturer of baked goods. We wanted to see if the quality of the packaging could be enhanced, made more attention-getting, by widening the border of the brand-name logo by one-quarter of an inch. A complex scheme was devised in which people were shown slides of new and old packaging in quick succession. If they noticed the new border, we knew it would be a success. Having no idea of what to expect, I was sent to shopping malls to grab shoppers and ask them to look at slides of a cookie package! It was an interesting introduction to research!

The initial excitement of gathering data tends to wear off once you learn basic research techniques. The process is the same whatever product you are investigating. The challenge lies in the analysis, not the actual research.

After a year at Crossely, I moved to AMF, another research company, as project director. I started dealing with clients, usually counterparts at my level; that is, people managing, rather than planning, a study. In mid-1982, I joined Louis Harris Associates as a research associate and was named research director in early 1983. I'm still overseeing detail work—that's something you really can't avoid even when you reach the level of vice president—but I have more creative input. At the end of a study I analyze data, meet with clients to go over results, and write a final report.

I now do a lot of work with banks. The creation of the "financial supermarket" has make the banking industry more conscious of the marketplace. As a result, I've had to learn a lot about finance. Banking is one area where research is just beginning to make inroads, and it's exciting to sell potential clients on the importance of research itself, not just the result of a particular study.

I like the variety that comes from working with a research supplier. You have to be a good juggler because you may be dealing with seven or eight projects at once. I like sitting down with seemingly unrelated data and creating a logical picture from them. But this analysis is interspersed with going out to meet people and sell them on a survey idea. It's not all desk work and it's not all sales; it's a combination of the two. Research provides a chance to use different parts of myself in different ways.

**John DeBiassio, Age 35**
**Vice President**
**Russell Research**
**New York, NY**

Although I've worked in market research for seven years, my association with it goes back much farther. Before I came to

Russell Research I worked in straight marketing with Progresso Foods. I started at Progresso, working in product management, after getting a master's in business administration. Eventually I became marketing services manager, and in this position I began to get involved with research. The last position I held at Progresso was assistant director of marketing; my functions included being research director. We never conducted any research of our own, but I worked with the research companies who researched the market for us.

I made the switch to market research for a number of reasons. Most of all, I find research fascinating. I also like the fast pace of a research company; it's much more active than straight marketing. The diversity of products to be dealt with was another strong attraction.

My marketing experience has been a valuable factor in my success in research. I came to this profession with a pragmatic approach that resulted from understanding marketing needs. I see beyond the research techniques. This results-oriented viewpoint improves my dealings with clients; I've been in their place and understand their goals.

I've worked on a wide variety of projects, from the simple to the complex. I've tested very specific promotional and advertising materials, such as a tag on clothing. Small details can make a difference in the way the public responds to a product. You must ask: what is the effectiveness of a colored tag versus a black and white one? Of one brand name over another? I've also handled much broader problems—evaluating entire marketing strategies, determining the best packaging, repositioning product lines. Here the questions require more analysis. Does the product live up to consumer expectations? Can it be sold to a specific segment of the market?

A complex project I've worked on was a tourism study for a country that hopes to attract American tourists. The study involved a series of questions: How do Americans perceive this country? Does the country have an image problem in the United States? If so, is it a major problem? How can it be overcome?

Often the results of a particular study show there is a problem—in packaging, advertising, marketing, whatever. Unfortunately, not all clients take this news and our advice as well as we would like. Some will request further research to confirm our findings. Of course, this in never the case when the study has favorable results!

# SECURITIES

FORTY million people and institutions in the United States alone trade in stocks and bonds. A report on the day's trading, the Dow-Jones Index, is a staple of the evening news broadcast. The business of selling securities is a large and lucrative one; top brokers and analysts earn as much as $400,000 a year. It's easy to see, then, why so many success-oriented people set their sights on "Wall Street"—a term that long ago came to mean the securities industry as a whole.

The term, of course, derives from the handsome, beaux arts-style building at 11 Wall Street that is the home of the New York Stock Exchange and is linked by half a million miles of telephone and telegraph wires to brokerage offices around the world. This vast communications network enables a buyer in London, England, for example, to purchase stock from a seller in California in a matter of minutes.

There are 6,935 brokerage firms in the United States registered with the Securities and Exchange Commission. Ranging in size from small, two-room operations to multinational giants like Merrill Lynch, they can be found in cities across the country and around the world. New York, NY, however, is the undisputed capital of the securities industry, offering more job opportunities than anyplace else.

But although the New York Stock Exchange, or "Big Board," is by far the largest central marketplace in the United States for securities trading, it's not the only one. The American Exchange is located nearby, and there are several regional exchanges: the Pacific, in San Francisco, CA, the Midwest, in Chicago, IL, and others in Boston, MA, Cincinnati, OH, and Philadelphia, PA.

Not every stock can qualify to be listed on one or more of the country's exchanges (requirements for the New York exchange are the most rigorous). Stocks are traded by brokers who are members of the exchange on which the stock is listed. (A brokerage house may be—and often is—a member of more than one exchange.)

When a broker receives a call from a client who wishes to buy a particular stock, the purchase order is directed to the floor of the appropriate exchange via computer. The brokerage firm has a representative there, called a floor broker. Every listed stock has a trading post, which is a specific location on the exchange's floor, and the floor broker goes there to ask for a "quote"—both the highest open bid made for the stock and the lowest available offer. Based on the quote, the broker offers a price, shouting out his or her bid for the number of shares the client wants. A floor broker with shares of that stock to sell calls out an offer to sell at the offered price, and a trade is made. The transaction is recorded immediately and the price of the stock is sent back to the broker's office by computer; the broker in turn relays the information to the client. The order is also sent over the wires and appears on the ticker tape in the office of every firm with a seat on that exchange.

Unlisted stocks are traded "over the counter." The broker can call up on an electronic visual display unit information listing the securities firms that trade in the various unlisted stocks, and the trade is then conducted directly by telephone.

Securities firms are currently locked in fierce competition with banks for customers' dollars. Recent federal deregulation permits both of these industries to offer products and services that were once the exclusive domain of the other. This means that a broker, in addition to selling stocks and bonds, may now offer clients an array of such products and services as asset management accounts, which pool all a client's assets into a single account. Checks may then be written against the consolidated account. The intense competition between brokers and bankers has resulted in converting many clients who were simple savers into investors.

Computers are revolutionizing the securities industry: speeding information from analysts to brokers, making possible complex computations that take some of the guesswork out of forecasting, and providing a host of other services that have eliminated much drudgery for the research department and enabled brokers to expedite their clients' orders. Analysts are using microcomputers to analyze balance sheets and cash flow and give brokers fast answers to clients' questions.

Now when a client calls a broker to learn the import of a company's $15 million increase in sales, for example, the analyst can punch the figure into the computer and be back to the broker in seconds with the answer. With the time the computer saves, countless more calculations can be performed in a day. Analysts can also look farther into the future, forecasting earnings three or four years ahead by trying out many different scenarios on the computer, whereas previously earnings were forecast for only a year ahead. And analysts are using the computer's word processing capabilities to produce research reports, complete with computer-drawn graphics, that can go right to the printer, eliminating the need for a typist or artist. When a stock is faltering, the broker can use the word processing capability to write individual letters to every client who holds the stock, recommending a call to the broker to discuss alternative investments.

Virtually every broker, no matter what the size of the firm, uses a microcomputer to store information on a client's holdings. This data may be cross-referenced in a variety of useful ways: by stock, by industry, or by investment objective, for example. Some firms are creating computer programs relating to options and bond trading that will balance the potential risks and gains in a proposed deal for a particular client.

Job categories in the securities field are:

- **Sales**
- **Research**
- **Operations**

The most prestigious and best-paying jobs are in sales and research. Sales requires high energy, excellent judgment, and, increasingly, some selling experience, and research requires sharp analytical skills (and more often than not an M.B.A.). Because the failure rate for recent college graduates entering the securities sales field with no relevant experience is extraordinarily high (some studies say it runs close to 95 percent) many firms now will only hire sales trainees who are at least 26 years old and have had some sales experience.

If you lack an M.B.A. degree or sales experience but are bent on a career in securities, an alternative for you might be the operations department—the "back office," where all the firm's transactions are processed. When sales slots open up, many firms fill them with the best and brightest from their operations department.

### Job Outlook

*Job Openings Will Grow:*   Faster than average

*Competition for Jobs:*   Keen

There are about 15,000 security analysts in the United States, compared with more than 80,000 brokers (and their ranks keep growing), making sales a considerably easier area to crack— assuming you have the required sales experience.

*New Job Opportunities:*The deregulation of the banking industry, the peak performances of the stock market in recent times, and the dizzying array of new options being made available to investors are creating new job opportunities in both sales and research.

Deregulation has paved the way for companies to buy up a variety of financial services and bring them all under one roof to create a "financial supermarket." The first to do so was Sears, Roebuck & Company: at more than 125 Sears store locations, customers can now make deposits at Allstate Savings & Loans, purchase insurance at Allstate Insurance, buy real estate through Coldwell Banker, and purchase securities through Dean Witter Reynolds. More than 850 new brokers were hired in 1983 to work in these Sears store locations, and brokers will continue to be added as more Sears stores open financial supermarkets. J.C. Penney has since followed Sears's lead.

The fierce competition for investors' money is spurring brokerage houses to enlarge their sales and research departments to attract customers, creating more jobs for brokers and analysts.

## Geographic Job Index

New York, NY, has the highest concentration of brokerage firms of any city in the United States, and most of the major firms are headquartered there, so it's the best place to find jobs in research and operations. Other cities with a high concentration of brokerage firms include Boston, MA, Philadelphia, PA, Chicago, IL, Dallas, TX, San Francisco, CA, and Los Angeles, CA. The major firms have an extensive network of branch offices (Dean Witter Reynolds, for example, has more than 325 branch offices throughout the 50 states), and even small cities have one or more brokerage offices. So you could find a job in sales almost anywhere in the country, although the field is larger, naturally, in a big city.

## Who the Employers Are

NATIONAL BROKERAGE FIRMS   employ thousands of employees in their nationwide branch offices. The biggest of them all, Merrill Lynch, employs more than 15,000 people. These firms maintain large research departments and spend millions of dollars each year tracking down the most attractive investments for their customers.

REGIONAL BROKERAGE FIRMS provide many of the same services offered by national firms but specialize in trading and promoting the interests of local companies. They employ fewer people than national firms in their offices (all of which are concentrated in their immediate area). Some very small brokerage firms have one office in one city only.

DISCOUNT BROKERAGE HOUSES are firms that do nothing but execute trades. They do not maintain research departments or offer advice, and their fees to investors are correspondingly lower. Many banks are forming partnerships with discount brokers so they can offer their customers discount brokerage services. Two such partnerships are Bank of America and Charles Schwab and Chase Manhattan Bank and Rose & Company. Many discount brokerage houses are national, but there are local ones as well.

COMMERCIAL BANKS have clients that are principally institutions and individuals with large sums to invest. They employ portfolio managers to handle such investments. Their staffs also include buy-side analysts, who offer purchase recommendations. (Unlike sell-side analysts, who are at securities firms that sell stock, buy-side analysts work for institutions making stock purchases.)

INSURANCE COMPANIES hire buy-side analysts also, who are responsible for advising the company about investing the huge sums of money collected as premiums from policyholders.

*Major Employers*

Allen & Company, New York, NY
Bear Stearns & Company, New York, NY
A.G. Becker, Inc., New York, NY
Dean Witter Reynolds, Inc., New York, NY
Donaldson Lufkin & Jenrette, Inc., New York, NY
Drexel Burnham Lambert, Inc., New York, NY
A.G. Edwards & Sons, St. Louis, MO
The First Boston Corporation, New York, NY
Goldman Sachs, New York, NY
E.F. Hutton Company, New York, NY
Kidder Peabody Company, Inc., New York, NY
Merrill Lynch Pierce Fenner & Smith, Inc., New York, NY
Morgan Stanley & Company, New York, NY
Paine Webber, New York, NY
Prudential-Bache Securities Brokers, New York, NY
L.F. Rothschild Unterberg Towbin, New York, NY
Salomon Brothers, Inc., New York, NY
The Securities Groups, New York, NY
Shearson Lehman/American Express, Inc., New York, NY
Shelby Cullom Davis & Company, New York, NY
Smith Barney Harris Upham & Company, Inc., New York, NY
Spear Leeds & Kellogg, New York, NY
Stephens, Inc., Little Rock, AR
Thomson McKinnon Securities, New York, NY

## How to Break into the Field

Because the securities industry is one in which the professionals tend to know their colleagues, the surest route to a job interview is through personal contacts—your school's alumni, family friends, neighbors, a relative's stockbroker. Failing that, try a letter-writing campaign: If you'd like a job as a broker write to the account executive recruitment office at the headquarters or to the branch manager at locations in your area. If your interest lies in operations, write to the operations manager at the firm's headquarters. Send a carefully worded letter stating your qualifications and requesting an interview. Enclose your résumé. Follow it up with a phone call requesting an appointment for an interview.

### International Job Opportunities

Extremely limited. Most of the major firms have offices abroad, but they tend to hire local residents for the positions that exist there.

# SALES

Brokers (also known as account executives, registered representatives, or salespeople) act as agents for people buying or selling securities. Because the performance of an account executive is crucial to the client's satisfaction and the firm's reputation, candidates are put through a rugged qualifying process at any large brokerage firm. The first hurdle is usually a general aptitude test; if you complete that successfully, you'll be interviewed by a succession of people, usually beginning with a corporate recruiter or a branch manager, who will rate your potential for success as a broker. The final hurdle will be a measurement of your sales skills in a test that includes exercises simulating problems and situations commonly faced by brokers. These involve telephone calls to prospects and relevant analytical work. Try to talk to a broker beforehand to prepare for this phase of the process.

As a beginning broker, your aim will be to build up a clientele. The best place to start is with people you know—family, friends, neighbors, members of groups or clubs to which you belong. You'll also be combing phone directories and mailing lists for names of prospective clients and spending the bulk of each day soliciting (many firms expect new brokers to make between 50 and

100 phone calls a day). While you continue to search for new business you'll be servicing your clients: keeping them abreast of their stocks' performance, executing trades, and recommending financial investments suitable to their needs and objectives.

Brokers usually specialize in one type of security, either stocks or bonds, and also either in retail sales, where your clients are individuals, or institutional sales. In addition, there are floor brokers, who work on the floor of a stock exchange, executing the actual trades of listed stocks.

To become a broker, you must pass the licensing exam given by the New York Stock Exchange, the main regulatory body for all the exchanges; in order to take the test, you must be sponsored by a firm. The firm that hires you will put you through an intensive account executive training program and give you study guides to help you prepare for the licensing exam. During the first three months of your training, while you prepare for the exam, you'll be observing the activities of a working brokerage firm. An additional month might be spent taking courses at the firm's training center.

During the next year, you'll be a broker-in-training at a branch office, with the manager of the branch serving as your supervisor. While you're in training, you'll be paid a salary. Once the training period ends, however, you'll be working strictly on commission, so your income will depend on how many transactions you process. Because brokerage firms demand a high level of productivity (many expect to see brokers earn about $50,000 in gross commission the first year), be prepared to work hard.

## Qualifications

*Personal:* Self-confidence. Personality. Foresight. Drive. Persistence. Ability to influence others. The strength to withstand frequent rejection.

*Professional:* Ability to work comfortably with numbers. Understanding of basic business concepts. Previous sales experience preferred.

## Career Paths

| LEVEL | JOB TITLE | EXPERIENCE NEEDED |
|-------|-----------|-------------------|
| Entry | Sales trainee | College degree, sales experience helpful |
| 2 | Account executive | 1-3 years |
| 3 | Branch manager | 4-6 years |
| 4 | Regional manager | 6-10 years |
| 5 | National sales manager | 10+ years |

## Job Responsibilities

### Entry Level

**THE BASICS:**   Identifying prospective clients through mailing lists and phone directories and making cold telephone solicitations. Answering clients' telephone queries. Reading financial publications. Processing transactions.

**MORE CHALLENGING DUTIES:**   Advising clients on appropriate investment strategies. Keeping current clients informed of their stocks' performance by telephone or letter. Studying reports from the research department.

### Moving Up

Your success depends on how hard you're willing to work—the number and quality of clients you can attract, your investment acumen, the soundness of the judgments you make on the basis of factual material from the research department, and your wiliingness to do more than simply take orders from your clients. It takes years to build a reputation as a broker who knows his or her business thoroughly. If you become a top performer in your branch and have managerial know-how, you may be offered the job of branch manager.

If you reach that point, you'll be required to relinquish all but a few of your clients. (You may be able to retain those with whom you have a personal relationship or who are your biggest investors.) As a branch manager, you'll be paid a salary plus a bonus based on the amount of money the branch office brings in. In addition, you will collect commissions on any transactions you continue to make.

# RESEARCH

A broker is only as successful as the company's research department. Knowing which stocks to go after and which to sell comes from listening to the presentations and reading the reports of the firm's researchers, or security analysts, who study stocks and bonds, assess their current value, and forecast their earning potential. Security analysts tend to specialize in a single industry, such as oil or steel, and quickly becoming experts in their area. Most security analyst jobs go to candidates who have an M.B.A. However, some brokerage firms will hire recent college graduates on the condition that they attend business school at night, and will offer tuition reimbursement.

## Qualifications

*Personal:* Ability to work under pressure. Foresight. Self-confidence. Ability to trust your own instincts.

*Professional:* Verbal and writing skills. Keen analytical skills. Familiarity with accounting procedures. Ability to read between the lines of annual reports. Facility working with a software "calc" program.

## Career Paths

| LEVEL | JOB TITLE | EXPERIENCE NEEDED |
|-------|-----------|-------------------|
| Entry | Research assistant/ junior analyst | M.B.A. degree helpful |
| 2 | Senior analyst | 3+ years |
| 3 | Managing director | 10 years |

## Job Responsibilities

### Entry Level

**THE BASICS:**  Reading financial reports. Analyzing corporate balance sheets. Working with figures. Making written recommendations to more senior analysts. Assisting senior people at whatever research work needs to be done.

**MORE CHALLENGING DUTIES:**   Accompanying senior analysts on visits to corporation officials to gather firsthand information about the company. Advising the firm's brokers on specific stocks. Fielding questions posed by brokers.

### Moving Up

After three or more years of gaining familiarity with and expertise in a particular industry, if you demonstrate that your analyses and interpretations of trends and developments are sound, you may be promoted to senior analyst. As a senior analyst, you'll be called on to answer any difficult questions posed by brokers or their clients, and act as adviser on all stocks related to the industry in which you are expert. You'll periodically visit branch offices to deliver oral presentations, accompanied by written reports, on your industry to brokers there. You'll also be accompanying institutional salespeople on their visits to lucrative accounts.

# OPERATIONS

The operations department, or "back office," is where the hundreds of thousands of daily transactions made by the firm's brokers are processed and recorded. The work is divided among several groups of clerks, each group with specific responsibilities. The purchasing and sales clerks make sure that every buy matches up with a sale by studying the computer printouts that record all transactions. The main source of this information is the Securities Industry Automation Corporation, an automated clearinghouse that is jointly owned by the New York and American exchanges. The printouts show every buy and sale on these exchanges in a single day; in addition, this same source provides information on national trading. Firms trading on exchanges outside New York, NY, receive comparable information from other automated sources. If the firm has made a buy and no corresponding sale appears on the printout, then a recording error has been made, and the purchasing and sales clerks must call around to other securities firms to try to learn who made the sale. Clerks in client services post dividends to clients' accounts and mail out monthly statements. Margin clerks keep track of clients' accounts, making sure they haven't purchased more on credit than is legally allowed. Compliance clerks ensure that transactions are completed according to all rules and regulations spelled out in the *New York Stock Exchange Constitution and Rules* book. Department heads oversee each of these services.

Securities are received and stored or transferred in a top-security area called the cage, where only a few people in the firm are allowed to enter. Cage clerks microfilm all securities and box them for storage in the vault or transfer them elsewhere to be stored.

## Qualifications

*Personal:*　Conscientiousness. Good powers of concentration. An eye for detail.

*Professional:*　An understanding of basic business concepts. An affinity for and ability to deal with numbers.

## Career Paths

| LEVEL | JOB TITLE | EXPERIENCE NEEDED |
|---|---|---|
| Entry | Clerk | College degree helpful |
| 2 | Supervisor | 2-3 years or an M.B.A. degree |
| 3 | Section head | 4-5 years |
| 4 | Manager | 6+ years |

## Job Responsibilities

### Entry Level

**THE BASICS:**   Checking printouts. Balancing sales and buys. Answering telephones. Microfilming and storing stock certificates. Mailing out monthly statements. Filing.

**MORE CHALLENGING DUTIES:**   Fielding brokers' questions. Managing a greater workload as your speed and efficiency increase.

### Moving Up

An accomplished clerk may be promoted to a supervisory position. As a supervisor you're typically responsible for five to seven clerks, assigning their work, monitoring their productivity, and offering guidance when needed. You will also do some administrative work, such as submitting absentee reports, preparing productivity reports for upper management, and establishing vacation schedules. A section head acts as liaison between departments and is prepared to deal with problems that may arise in day-to-day operation so that the department functions smoothly and effectively.

# ADDITIONAL INFORMATION

## Salaries

SALES commissions vary with the type of security and the size of the transaction. Retail brokers collect between 30 and 40 percent of the fee that the firm charges for each transaction; institutional brokers collect somewhat less—around 15 percent—because large blocks of securities are being traded. For brokers who bring in a high volume of business, there are numerous incentives, such as free trips and raises in commission. The income potential is unlimited, and some brokers gross in excess of $1 million a year in commissions.

RESEARCH salaries range from $30,000 a year for entry-level M.B.A.s to six-figure incomes for managing directors. The analysts are rated by the brokers on the basis of the quality and depth of their research and their record of success. These ratings are carefully considered when analysts are up for their biannual bonuses.

OPERATIONS salaries for clerks range from $9,000 to $20,000 a year, plus overtime, depending on experience. Salaries for supervisors with M.B.A. degrees start at $30,000 a year, plus bonuses.

## Working Conditions

*Hours:*    A broker's day usually begins at 8 A.M., in time to read the papers and financial journals and talk with the research department before the exchanges open at nine. Brokers often leave the office once trading ends at 4 P.M. Operations work is usually nine to five, with overtime when trading is heavy. Supervisors and section heads put in slightly longer hours, perhaps eight-thirty to six, to catch up with administrative details or attend meetings. Research analysts work the longest hours, typically past 7 P.M.

*Environment:* Junior brokers, clerks, and junior analysts work in bullpen arrangements. Senior brokers, operations managers, and senior analysts have private offices. Typically, those in sales enjoy the plushest surroundings.

*Workstyle:* Sales personnel spend a great deal of time on the phone, either speaking with established clients or soliciting new business. Institutional salespeople may wine and dine big clients after normal business hours. Clerks and operations supervisors spend nearly all their time on paperwork, but operations managers may be in meetings up to half of each day. Research is also a desk job; the study of financial statements involves using the microcomputer to arrive at various indicators of a company's financial status: asset/debt ratio, sales/inventory ratio, sales/debt ratio. You will meet at least once a week with other members of your research team.

*Travel:* Opportunities to travel are nonexistent for brokers and operations staffs. Regional and national sales managers visit branch offices frequently. As a research analyst, how much you travel and how far you go depends on the industry you cover. If you specialize in an industry the center of which is in your home area, out-of-town trips may be infrequent.

### Extracurricular Activities/Work Experience

Team sports—participating as a member or leader

Investment clubs—participating as a member

Sales experience—working in sales of any kind; telephone solicitation work a plus

### Internships

Internships are not easily arranged in the securities industry. Your best bet might be to apply to small, local brokers; however, any prospective sponsor of an internship will expect to see a solid academic record.

## Recommended Reading

**BOOKS**

*The Money Game* by Adam Smith, Vintage: 1976

*The Money Messiahs* by Norman King, Coward-McCann: 1983

*Stealing from the Rich: The Story of the Swindle of the Century* by David McClintick, Quill Publications: 1983

**DIRECTORIES**

*Broker-Dealer Directory* (annual), Securities and Exchange Commission, Washington, DC

*Security Dealers of North America* (semiannual), Standard & Poor's, New York, NY

*Who's Who in the Securities Industry* (annual), Economist Publishing Company, Chicago, IL

**PERIODICALS**

*Barron's* (weekly), Dow Jones & Company, 22 Cortlandt Street, New York, NY 10007

*Business Week* (weekly), McGraw-Hill, Inc., 1221 Avenue of the Americas, New York, NY 10020

*Dun's Business Month* (monthly), Dun & Bradstreet Corporation, 875 Third Avenue, New York, NY 10022

*Financial World* (bimonthly), Macro Communications, Inc., 1250 Broadway, New York, NY 10001

*Forbes* (biweekly), 60 Fifth Avenue, New York, NY 10011

*Fortune* (biweekly), Time-Life Building, Rockefeller Center, New York, NY 10020

*Institutional Investor* (monthly), 488 Madison Avenue, New York, NY 10022

*Investment Dealers Digest* (weekly), 150 Broadway, New York, NY 10038

*Money* (monthly), Time-Life Building, Rockefeller Center, New York, NY 10020

*The Wall Street Journal* (daily), Dow Jones & Company, 22 Cortlandt Street, New York, NY 10007

*Weekly Bond Buyer* (weekly), 1 State Street Plaza, New York, NY 10004

## Professional Associations

Financial Analyst Federation
1633 Broadway
New York, NY 10019

National Association of Security Dealers
2 World Trade Center
New York, NY 10048

Securities Industry Association
120 Broadway
New York, NY 10271

# INTERVIEWS

**Assistant Vice President, Research
L.F. Rothschild Unterberg Tobin, New York, NY**

I was a triple in college—history, political science, and economics—and I hasten to add that I managed to do that major in economics without taking any courses in mathematics. I definitely did fall into that category of women who have a negative reaction to numbers. Upon graduation I worked for the Corporation for Public Broadcasting in their human resources development. Part of my responsibility was monitoring the employment and portrayal of women and minorities in public broadcasting. That necessitated doing quarterly reports to a congressional committee, and I had to start compiling employment figures, statistics—and there was math staring at me! I later became responsible for the departmental budget, which was rather substantial as our department was responsible for handing out training grants throughout the system. I quickly discovered there was nothing to be frightened about.

I enjoyed that for a while, but got somewhat tired of the nonprofit orientation. It's not very insightful, but the way I ended up in securities was to look in Washington, DC, and to find out what kind of private oriented enterprises there were. And there were not many. I won't say "securities" bounced right out of the phone book—but it was the only industry I could see getting into without a great deal of difficulty. I had invested with some success on my own and found it interesting, so I investigated the various brokerage firms in Washington, and found that Ferris and Company, which is a fine regional house in Washington, had a superior, intensive training program.

Once I was in securities I found out that if I wanted to go beyond the basic retail broker status, I had to have a graduate degree. And that's why I went back to school to get an M.B.A. I went to George Washington University while I was still working at Ferris. Because the only management position available at a regional house would be a branch manager, and that certainly was out of the question with my few years of experience, and not having built and enormous clientele book, I decided to come back to New York, which is my native state.

I never liked pure sales and never did cold calling, and quite frankly was uncomfortable with pure commission as a source of income. So I found myself getting more and more involved in the total financial picture of my clients, which gave me a greater level of security in terms of what I was or was not doing with their money. And that was quite suitable experience for the posiion I now hold in Rothschild's research department. I'm in what's known as portfolio research, and that job entails essentially being a broker to our brokers. They submit their client portfolio with the appropriate investment objective information, and we analyze the portfolio or develop a portfolio to meet the client's needs. The beauty of this is that it gives the client excellent service, because this is done at no additonal fee, and I have no vested interest in whether the broker buys or sells. I make decisions on a needs basis for the client versus a need basis for the broker.

The bottom line about having the M.B.A. is that it helped me get the job. I don't know that I use more that 10 percent of what I learned. The program that I was involved in was more qualitative than quantitative, which I frankly liked, and I think an awful lot of large corporations are coming to the conclusion that the quantitative programs are great in the short term, but they're finding that the long-term objectives of many corporations are being sacrificed. That's a consideration one should look at seriously when picking an M.B.A. program. I also think one should work before going for a graduate degree. Although there are hordes of recruiters on campuses these days, and M.B.A.s are still pulling in a fairly nice salary for an initial job, I do think that the allure of the degree without work experience is rapidly dissipating. More and more employers are saying that without experience what you've learned means nothing to you because you haven't been able to apply it while you learned it. The M.B.A. was more a premium degree when I got it than it is today, but it's a question of having the degree to get the door open now.

Sixty to 70 percent of my day is spent talking to brokers, responding to questions on particular stocks about whether they're appropriate for specific clients, looking at portfolios, and going to meetings with other analysts to look at companies or to discuss the general market outlook. The balance of the time is spent with my clients, which is the icing on the cake. I enjoy my salary compensation, and then can do as much commission business as I want. I'm constantly reading research reports of other firms or independent research organizations, and company annual reports, and doing spread sheets on earnings projections. I also function as a conduit between the specialized analysts who cover specific industries and the broker for those securities that our firms covers on a regular basis. When we're talking about companies that are not regularly followed, that's when I have to look at them.

I like the sense of power in this job. It's really rather heady to have brokers who've been on Wall Street for 30 years have to ask me if it's okay to by or sell something. But I also enjoy the research end. I have found—much to my surprise over the years—that numbers are not intimidating at all. It gives me the opportunity to be both a broker for those clients that I do handle without being compelled to trade in their accounts to make my living. And I like being of assistance to the brokers because, although I don't have a vested interest in whether they buy or sell, over the long term if you do a portfolio structure that is appropriate for their client they keep the client. It's not a question of perhaps buying one or two stocks that don't turn out so the client goes to some other broker. If you can do a total picture so that no single security will make or break them, they're going to keep their portfolio with you.

**Leah Pfeffer**
**Administrative Manager, Institutional Sales**
**Dean Witter Reynolds, Inc., New York, NY**

My first profession was teaching—I have a B.S. and an M.S. in education—but after three years at the head of a third-grade classroom, I was ready for something new. I was interested in business (and I must admit I was ready to work with adults), but beyond that I didn't have a clear idea of where to start looking. An employment agency sent me to interview for a position as a sales assistant with a brokerage house. I knew little about either securities or sales, but the job interested me and it met my two basic requirements—it didn't require typing and I wouldn't be taking a cut in pay by switching jobs. I didn't get that job, but I found what I had been looking for. I applied for other sales assistant openings and wound up at Thomson McKinnon. That was in 1968.

I learned a great deal about securities. The firm sent me to the New York Institute of Finance, which prepared me to take (and pass) the registered representative exam with the New York Stock Exchange. But I knew I didn't want to be an assistant forever. I pursued sales, moving to a small brokerage house called Hirsch and Company. Few woman were in sales at that time. In fact, I have wondered if the primary reason I got that job was because I was interviewed by one ot the few female partners then in the business.

I received no formal training—I was given a phone and a desk, and I was on my own! Building a client base was tough. I found myself in a bear market with few products to sell. At the time, brokers dealt mainly with stocks. We sold some bonds, but the options market was really just starting. Today, a broker has much more to work with.

The bad part of being a broker is that you are always on the job. Wherever you go, whomever you meet, one thing is foremost in your mind—making client contacts. I did a lot of cold calling. I did hit on a trick to make contacts, however. I would go through the phone book, calling everyone with the name of Pfeffer. By playing up the coincidence of our names, I could break the ice, and often people would talk to me because of the connection.

The firm went out of business, which gave me a chance to reevaluate my goals. I came to the conclusion that I would be happier not selling. I was not bad at selling, but the job just didn't fit my personality. I went back to Loeb Rhoades (now a part of Shearson Lehman/American Express), handling day-to-day, administrative details as a supervisor. After four years, I moved to Bache Halsey Stuart, Inc. (now Prudential Bache), where I spent another four years as manager of marketing and support services in the institutional sales department.

In institutional sales, we sell our product—our research—to major clients, such as bank trust departments and large pension funds. I was responsible for discovering what our clients needed in terms of the research itself and what they expected in terms of its presentation. Some wanted a broad analysis; others asked for more specific information. The data must be easily understandable and, above all, must be timely.

I left Bache after four years, moving to Dean Witter Reynolds. I still work with institutional sales, but am more involved with overseeing department functions. Institutional sales is now getting into new areas. Electronic transmittal of data is speeding our delivery to customers. And we are now looking into closed circuit television. With it, we will be able to contact our clients directly; our research analysts and salespeople won't travel, but may make all their analytical presentations and sales calls on television. Your clients always expect you to have a crystal ball. Not so long ago, clients wanted forecasts for the coming year or two; now they want predictions five years in advance! As research techniques become more sophisticated, our forecasts seem to be getting much more accurate, but of course room for error remains.

Although I do not do actual selling, my sales background has helped me immensely. I understand the pressures on our salespeople. And in a sense, I still do some selling—not to the clients, but to the salespeople. I tell them what we can supply to clients, and I motivate them to sell our services.

I am a member of the Financial Women's Association of New York, an organization that includes women from many different financially oriented professions. I enjoy meeting other professional women to compare notes and exchange information. And, as you progress in your career, networking is important.

I have also supplemented work experience with an M.B.A. I went at night, taking six years to complete the degree. As you can tell, I wasn't in a hurry! The best place to learn is on the job, but I felt I needed the M.B.A. to remain competitive. So far, having the degree has not made a difference in my job; however, if I decide to look for another job, it may be valuable. The M.B.A. is simply becoming more common. In order to compete against other advanced-degree holders, you need it.

# BIBLIOGRAPHY

*The College Graduate's Career Guide* by Robert Ginn, Jr., Charles Scribner's Sons: 1981

*College Placement Annual* by the College Placement Council: revised annually (available in most campus placement offices)

*The Complete Job-Search Handbook: All the Skills You Need to Get Any Job and Have a Good Time Doing It* by Howard Figler, Holt, Rinehart & Winston: 1981

*Consider Your Options: Business Opportunities for Liberal Arts Graduates* by Christine A. Gould, Association of American Colleges: 1983 (free)

*Go Hire Yourself an Employer* by Richard K. Irish, Doubleday & Company: 1977

*The Hidden Job Market for the 80's* by Tom Jackson and Davidyne Mayleas, Times Books: 1981

*Jobs for English Majors and Other Smart People* by John L. Munschauer, Peterson's Guides: 1982

*Job Hunting with Employment Agencies* by Eve Gowdey, Barron's Educational Series: 1978

*Making It Big in the City* by Peggy J. Schmidt, Coward-McCann: 1983

*Making It on Your First Job* by Peggy J. Schmidt, Avon Books: 1981

*National Directory of Addresses and Telephone Numbers*, Concord Reference Books: revised annually

*The National Job-Finding Guide* by Heinz Uhrich and J. Robert Connor, Doubleday & Company: 1981

*The Perfect Résumé* by Tom Jackson, Doubleday & Company: 1981

*Put Your Degree to Work: A Career Planning and Job Hunting Guide for the New Professional* by Marcia R. Fox, W.W. Norton: 1979

*The Student Entrepreneur's Guide* by Brett M. Kingston, Ten Speed Press: 1980

*What Color Is Your Parachute? A Practical Manual for Job Hunters and Career Changers* by Richard N. Bolles, Ten Speed Press: 1983

*Where Are the Jobs?* by John D. Erdlen and Donald H. Sweet, Harcourt Brace Jovanovich: 1982

# INDEX

Academic background: actuaries, 92, 98, 103; banking, 2; market research, 113

Adult education, 78

AIM (automated teller machines), 37

Allstate Insurance, 129

American Marketing Association, 118–119

Analyst, 116, 136, 140

Apple Computers Inc., 41, 48, 54

Applications intelligence (AI), 24–26, 33

Artificial intelligence (AI), 49

Atari (co.), 41

Banking, 1–21, 36, 113

Bonds, 127, 133, 135

Branch Managers, 132, 135

Brand managers, 118

Breaking into the field: banking, 4; computer industry, 33, 41, 49; department store retailing, 63; education, 79; insurance, 94; market research, 115

Broker, 126, 127, 128, 132, 140

Brokerage firm, 129, 130

Bureau of Labor Statistics, 96, 98

Business Research Practicum, 119

Buyer, 60, 65–68

Cable television, 61, 113

Cage clerks, 137

Caldwell Banker, 129

Career advancement and paths: banking, 7–12; computer industry, 26, 27, 28, 29, 30, 31, 32, 39, 46, 51: securities, 134, 136, 138, 139; market research, 117: insurance, 97, 99, 101, 102; department store retailing, 65–67; education, 81, 82

Casualty Actuarial Society, 99–100

Certified life underwriter (CLU), 95

Chartered financial planner (CHFP), 95

Chartered property casualty underwriter (C.P.U.C.), 95, 101

Chase Manhattan Bank, 8

Chemical Marketing Research Association, 115

CODASYL (hierarchical data base package), 28

Coder, 116–117

College of Insurance, 94

Commercial banks, 3, 5, 15, 130

Compliance clerks, 137

CAD (computer aided design), 35

CAI (computer aided instruction), 53

CAM (computer aided manufacture), 49

Computer-drawn graphics, 127

Computer industry, 23–58, 75, 77, 112, 126–127

Computer languages, 25, 33–34

Computer technicians, 37

Consumer goods manufacturer, 114

Continuing education, 77

Credit lending, 2, 3, 5, 14

Data analysts, 112

Data base management, 27

Data base specialists, 28

Data processing, 32

Dean Witter Reynolds, 129

Deere and Company, 35

Department store retailing, 59–73

Documentation specialist, 31

Dow-Jones Index, 125

Editor, 116, 117

EDP (electronic data processing), 30

EDP auditor, 30

Education, 75–89

Employers: banking, 3, 4; computer industry, 38, 41, 45, 57; department store retailing, 61, 62; education, 78; insurance, 93; market research,

114; securities, 130–131;
Engineers, 33
Entrepreneurs, 1
Entry-level positions: banking, 7, 9, 11, 13; computer industry, 39, 40, 41, 44, 46–47, 57; department store retailing, 60, 63–64; education, 82; insurance, 93, 97, 100, 102–104; marketing research, 113, 115, 117; securities, 132–133, 134, 136
Experience, prior: education, 81, 85; department store retailing, 65, 69; insurance, 97, 99, 101, 104; market research, 119; securities, 138, 140
Extracurricular activities: banking, 15; department store retailing, 69; insurance, 104; market research, 119; securities, 140
Equal Opportunity Commission, 92

Federal deregulation, 126, 127
Federal Job Information Center, 36
Floor broker, 126

Georgraphic job index, banking, 3; computer industry, 34, 42, 50; department store retailing, 61; insurance, 93; market research, 114; securities, 129

Hardware, 23, 25, 26, 28, 48
Honeywell, 48

IBM, 41, 48
Information analyst, 112
Information systems, 3
Internships: banking 5, 15; department store retailing, 69; education, 85; insurance, 105; market research, 119; securities, 140
Insurance, 91–110, 130
Insurance agents, 96
Insurance careers, 93
International job opportunities: banking, 5; department store retailing, 63; education, 80; insurance, 95; market research, 115;

Investment banking, 6
securities, 132
Investor, 129

Job outlook: banking, 3; computer industry, 23, 32, 41, 48; department store retailing, 60; education, 77; insurance, 92; market research, 113; securities, 128
Job responsibilities: banking, 6, 7–8, 12; computer industry, 26–32, 38, 46, 51; department store retailing, 64, 67; education, 81; insurance, 97, 100, 102; market research, 117; securities, 134, 136, 138

Licensing (insurance), 96
Life Underwriting Training Council, 102

Management positions; banking, 5, 8; computer industry, 30, 33, 57; department store retailing, 60, 61, 69; market research, 118
Manufacturers, 35, 68, 111
Margin clerks, 137
Market research, 111–124
Marketing, 29, 33, 49
Marketing services, 115
Marketing strategies, 112
Merchandising, 63, 65
Merrill Lynch, 126
Microcomputers, 1, 26, 29, 54, 127
Mitsubushi (Co.), 48
Multinational corporations, 3, 5

National Council of Accreditation of Teacher Education, 79
National Education Association, 83
National Placement Organization, 79–80
New York Stock Exchange, 125, 126, 133
New York Stock Exchange Constitution and Rules (book), 137

Operations, 1–3, 5, 8, 14–15, 128–129, 137–139

Operations managers, 140
Original Computer Camp, 53

Point-of-sale computer terminals, 60
Preschool education, 78
Principal, 82
Private banking, 3
Product support representatives, 28–29
Professional associations: banking, 16; department store retailing, 70; education, 87; insurance, 106; market research, 120; securities, 142
Programmers, 34, 37, 44, 49
Programming (systems), 23–26, 29–32, 36, 52
Property and casualty insurance, 91

Qualifications: banking, 6, 8, 10, 12; computer industry, 25, 27–31; department store retailing, 64, 66; education, 87; insurance, 96, 99, 101; market research, 116; security, 133, 135, 137

Real estate, 36, 129
Recommended reading: banking, 15; department store retailing, 70; education, 86; insurance, 105; market research,120; securities, 141
Regional banks, 4
Religious organizations, 78
Research, 112, 128–129, 135, 136, 139
Research analysts, 115–116
Robots, 35
ROM (read only memory), 50

Salaries: banking, 13; department store retailing, 67; education, 83; insurance, 102; market research, 118; security, 133, 139
Sales: computer industry, 29, 33, 40, 46, 50, 57; insurance, 92, 96–98, 102; securities, 128–129, 132–134, 139
Savings and loan associations, 4, 15

School administrations, 82
Sears Roebuck & Company, 129
Securities, 125–143
Securities and Exchange Commission, 125
Securities Industry Automation Corporation, 137
Security analysts, 128, 135
Silicon Valley, 42
Social Security Administration, 36
Society of Actuaries, 99
Software, 26, 28–29, 37, 41–42, 44–45, 50, 53
Special education, 77
Stocks, 126, 133, 135
Study and Teaching Opportunities Abroad (Pat Kern McIntyre), 81
Subliminal messages, 111
Superintendents, 82
Systems analysts, 23–24, 26, 28–29, 32, 36–37, 49
Systems (banking), 3, 5, 10, 14
Systems design, 33

Tabulation personnel, 112, 116
Teacher, 80, 84
Teacher certification, 79, 81, 85
Teacher Exchange Branch, 80
Teaching, 81; business, 83; math, 77; secondary school, 80
Technical support representatives, 28
Technical writer, 31, 41
Technology, 1, 10
Telecommunication, 2
Texas Instruments, 41
Training, in-house: banking, 2, 7, 9, 11–12, 14; computer industry, 26, 53; insurance, 93; securities, 133
Travel opportunities: banking, 15; computer industries, 47, 52; department store retailing, 69; education, 85; insurance, 104; market research, 119; securities, 140
Travelers Insurance Company, 91
Trusts, 3, 5, 12
Turnkey suppliers, 41–42, 44–45

Underwriters, 92, 101, 103
United States Department of Defense, 34
United States Department of Education, 77
United States Department of Labor, 23
United States Department of Labor's Bureau of Labor Statistics, 33
United States Information Agency, 80
Universal Product Code, 37
User-friendly computers, 54

Video retailing, 61
Visual display terminals (VDTs), 28

Wall Street, 125
Wang Laboratories, 48
Working conditions: banking, 14; computer industry, 40, 46; department store retailing, 68; education, 84; insurance, 104; market research, 118; securities, 139–140

# NOTES

**NOTES**

# NOTES

NOTES

**NOTES**

# NOTES

**NOTES**

# NOTES

**NOTES**

# NOTES

# NOTES

---

---

**NOTES**